The Chronologers' Quest

Episodes in the Search for the Age of the Earth

The debate over the age of the Earth has been going on for at least two thousand years, and has pitted astronomers against biologists, religious philosophers against geologists. *The Chronologers' Quest* tells the fascinating story of our attempts to determine a true age for our planet.

This book investigates the many methods used in the search: the biblical chronologies examined by James Ussher and John Lightfoot; the estimates of cooling times made by the Comte de Buffon and Lord Kelvin; and the more recent investigations of Arthur Holmes and Clair Patterson into radioactive dating of rocks and meteorites.

The Chronologers' Quest is a readable account of the measurement of geological time. Little scientific background is assumed, and the book will be of interest to lay readers and earth scientists alike.

PATRICK WYSE JACKSON is a lecturer in geology and curator of the Geological Museum in Trinity College Dublin, and is a member of the International Commission on the History of Geological Sciences.

EON	ERA	PERIOD	EPOCH	AGE
PHANEROZOIC	CENOZOIC	NEOGENE	HOLOCENE	0.012
			PLEISTOCENE	
				1.8
			PLIOCENE	
				5.3
			MIOCENE	
				23
		PALEOGENE	OLIGOCENE	
				34
			EOCENE	
				56
			PALEOCENE	
				65
	MESOZOIC	CRETACEOUS		
				146
		JURASSIC		
				200
		TRIASSIC		
				251
	PALAEOZOIC	PERMIAN		
				299
		CARBONIFEROUS	PENNSYLVANIAN	318
			MISSISSIPPIAN	359
		DEVONIAN		
				416
		SILURIAN		
				444
		ORDOVICIAN		
				488
		CAMBRIAN		
				542
	PROTEROZOIC			
				2,500
	ARCHEAN			
		ORIGIN OF THE EARTH		4,550

The Geological Column with the age in millions of years of the start of each major stratigraphical unit (simplified and modified from the International Stratigraphic Chart published in *Episodes* 27, part 2 (2004), 85).

The Chronologers' Quest

Episodes in the Search for the Age of the Earth

PATRICK WYSE JACKSON

Trinity College Dublin

CAMBRIDGE UNIVERSITY PRESS
Cambridge, New York, Melbourne, Madrid, Cape Town, Singapore, São Paulo

Cambridge University Press
The Edinburgh Building, Cambridge CB2 2RU, UK

Published in the United States of America by Cambridge University Press,
New York

www.cambridge.org
Information on this title: www.cambridge.org/9780521813327

First published 2006

Printed in the United Kingdom at the University Press, Cambridge

A catalogue record for this publication is available from the British Library

ISBN-13 978-0-521-81332-7 hardback
ISBN-10 0-521-81332-8 hardback

*To Vanessa in Dublin, and Marcus and Eric
in Carlisle, Pennsylvania*

Contents

List of illustrations   *page* viii

List of tables   x

Preface   xi

Acknowledgements   xvi

1 The ancients: early chronologies   1

2 Biblical calculations   13

3 Models of Aristotelian infinity and sacred theories
   of the Earth   32

4 Falling stones, salty oceans, and evaporating waters: early
   empirical measurements of the age of the Earth   47

5 Thinking in layers: early ideas in stratigraphy   66

6 An infinite and cyclical Earth and religious orthodoxy   86

7 The cooling Earth   105

8 Stratigraphical laws, uniformitarianism and the
   development of the geological column   119

9 'Formed stones' and their subsequent role in
   biostratigraphy and evolutionary theory   154

10 The hour-glass of accumulated or denuded sediments   171

11 Thermodynamics and the cooling Earth revisited   197

12 Oceanic salination reconsidered   210

13 Radioactivity: invisible geochronometers   227

14 The Universal problem and Duck Soup   250

Bibliography   261

Index   285

Illustrations

Frontispiece: The Geological Column.   *page* ii

1.1 Ptah, the Egyptian Creator God.   4

1.2 Tablet with cuneiform inscriptions from the library
      of Assur-bani-pal, King of Assyria.   5

2.1 *The computacion of the ages of the worlde* from *Cooper's
      Chronology* (1560).   16

2.2 James Ussher (1580–1656).   17

2.3 First page of James Ussher's *Annales veteris
      testamenti* (1650).   25

3.1 Thomas Burnet (1635–1715).   33

3.2 Title page of Thomas Burnet's *Telluris Theoria Sacra* (1681). 40

4.1 Edward Lhwyd (1660–1709).   48

4.2 Fossil cephalopods and gastropods from Edward Lhwyd's
      Lithophylacii Britannici Ichnographia (1699).   53

4.3 Llanberis Pass, North Wales.   56

4.4 Edmond Halley (1656–1742).   59

4.5 Title of Halley's 1715 paper on lacustrine salination. 61

5.1 Nicolaus Steno (1638–1686).   67

5.2 Steno's diagram showing the stages of the development
      of the Tuscan landscape.   68

5.3 John Strachey's cross-section through the Earth (1725). 76

5.4 Giovanni Arduino's geological cross-section through
      the Valle dell'Agno (1758).   80

6.1 James Hutton (1726–1797).   87

6.2 Title page of Hutton's 1785 abstract.   90

6.3 The Giant's Causeway, Co. Antrim, Ireland.   97

6.4 Junction of granite and schist at Glen Tilt, Scotland. 100

6.5 Unconformity at Siccar Point, Scotland.   102

7.1 Georges-Louis Leclerc, Comte de Buffon (1707–1788).   107

7.2 Map of the district around Montbard, France. 112
7.3 Buffon's forge at Montbard. 113
8.1 Adam Sedgwick (1785–1873). 120
8.2 Roderick Impey Murchison (1792–1871). 120
8.3 Charles Lyell (1795–1875). 121
8.4 William Smith (1769–1839). 127
8.5 The 'Temple of Serapis' near Naples. 132
9.1 Whitby snakestones: Jurassic ammonites
Dactylioceras commune. 159
9.2 Plate from William Smith's *Strata Identified
by Organized Fossils.* 161
9.3 Stratigraphical chart showing characteristic fossils (1896). 163
10.1 John Phillips (1800–1874). 177
10.2 Geological map of the Weald, southeast England. 181
10.3 Charles Darwin (1809–1882). 182
10.4 Charles Doolittle Walcott's map of the geology of the
American mid-West (1893). 191
11.1 William Thomson, Baron Kelvin of Largs (1824–1907). 198
11.2 Lord Kelvin's coat of arms. 207
12.1 John Joly (1857–1933). 212
12.2 The Oceans. 218
13.1 Ernest Rutherford (1871–1937). 230
13.2 Radioactive decay series for uranium, thorium
and actinium. 234
13.3 Advertisement from *Le Radium* offering radium,
polonium, actinium and uranium salts for sale (1908). 235
13.4 Arthur Holmes *c.* 1911. 238
13.5 Pleochroic halo in biotite in Leinster granite from
County Carlow, Ireland. 239
13.6 Holmes' 1947 geological timescale. 247
13.7 Seven isochrons indicating an age of 3,400 million years
for the Earth (1947). 248
14.1 Clair Cameron Patterson (1922–1995). 253
14.2 Canyon Diablo and Henbury meteorites. 256
14.3 Title and abstract of Clair Patterson's 1956 paper. 258

Tables

2.1 Thomas Allen's 1659 chronology from Creation
     to the death of Christ.   *page* 18
7.1 Results of Buffon's heating and cooling experiments
     on metal spheres of various diameters.   116
8.1 The authors of the various divisions of the geological
     column, and dates first used.   136
10.1 Various estimates of the age of the Earth derived by the
     sediment accumulation method.   188
13.1 Radioactive decay series most usually used for
     geological dating.   242
13.2 Changing views of Arthur Holmes' Geological
     Timescale 1911 to 1960, compared with the 2004
     timescale published by the International Commission
     on Stratigraphy.   245
14.1 Meteorites dated by Clair Patterson in his 1953
     and 1956 papers.   257

Preface

Geologists have been much censured
for vainly endeavouring to assign measures of time
to the seemingly vague and shadowy ages
of the Trilobites and Belemnites.

John Phillips (1800–1874), *Life on the Earth, its Origin and Succession* (1860)

Some drill and bore
The solid earth and from the strata there
Extract a register, by which we learn
That He who made it and reveal'd its date
To Moses, was mistaken in its age.

William Cowper (1731–1800), *The Task* (1785)

I have two main reasons for writing this book, and both have their origins in family matters. A few years ago I spent a fortnight with my wife and two young daughters on holiday on the Dingle Peninsula, in southwest Ireland. This area of immense scenic beauty and cultural significance is also an area of 'classic' geology. As an undergraduate student I had followed in the footsteps of geologists such as George Victor Du Noyer, a noted antiquarian and watercolourist, and Joseph Beete Juke, his boss in the Geological Survey of Ireland, in mapping some Silurian and Devonian sediments that formed the backbone of the peninsula. I doubt I produced a fuller and more accurate map than did these early pioneers. I recalled with feeling, during the first three damp, rain-sodden days of our holiday, the remark of Sir Roderick Impey Murchison, one-time Director of the Geological Survey of Great Britain, who declared, having endured two weeks of such weather, that 'there was nothing of interest in Irish geology'.

However, the changing weather conditions, allied with the splendid sunsets that we witnessed during the first week of our holiday, clearly left its mark on my elder daughter. She saw beautiful salmon-pink clouds streaking across the Kerry sky, which reflected the favourite culinary dish of my wife. She heard about the unusual 'green flash' that occasionally accompanied the very last vestiges of the orange sphere as it disappeared beneath the distant horizon – but was not fortunate enough to see it. This daily cycle of dawn, morning, afternoon and sunset got her thinking, and out of the blue as we crossed the mountainous road of the Connor Pass, a little voice from the back of the car asked, 'Mummy, how long ago did the world begin?' Quickly, realising my interest in the subject, my wife deflected the question to me.

'How long ago did the world begin?', I thought, pausing to reflect on the complexity and indeed simplicity of such a question from a person who had only celebrated her fifth birthday a month earlier. If I had attempted to fob her off with a response such as 'Oh, a long time ago' or 'Well, sometime before Granny was born', I knew that this would have been most unsatisfactory from the perspectives of both Susanna and myself. 'The world is over four thousand million years old,' I replied as I turned around. 'That's a lot of noughts, isn't it,' she thought out loud. And she was right, it is a lot. For a few moments she took this in, and appreciated that the world was very old indeed.

On my return to the city, I met up with my youngest brother Michael for our usual weekly lunchtime escape from respective offices and he handed me two items. One was a book and the other a large roll of paper. He was aware that I was beginning this book, and said, 'You're interested in James Ussher. Have a look at these.' The book was a small green octavo volume entitled *The Life and Times of Archbishop Ussher*, written by a Reverend J. A. Carr, Rector of Whitechurch, a small parish situated four miles south of Dublin that nestles on the northern slopes of the local granitic mountains. He had found it on a upper shelf in a bookcase in my mother's house, and I was delighted that he had, as I had been trying to track down a copy of Carr's book, perhaps the best and most accessible biographical treatment of the Archbishop published in the nineteenth century. The second item, the large roll of paper, proved to be of great personal interest, but unfortunately of less use to me here. I carefully unrolled the four-foot-long document on the table and saw it was a family tree. Right at the top was written 'Henry Jones = Margaret Ussher (sister of the Primate)'. I gazed at the multitude of names and dates, interconnected by a maze of straight and wavy lines, and passed my eyes over several generations. Another Henry Jones listed was Bishop of Meath between 1661 and 1682 and was responsible for rescuing the Book of Kells, the seventh century version of the Gospels, from a bog in County Meath. This, the finest of Irish illuminated manuscripts, is on show in the Library of Trinity College Dublin, where it is seen by nearly a million tourists each year. Another character by the splendid name of Rashleigh Belcher caught my eye. He was a medical doctor who practised in the market town of Bandon in County Cork. I finally made my way down to the bottom of the document and there in plain black ink was my name. Amazed, I turned to my brother and remarked, 'We're related by marriage to James Ussher!' and added with a laugh, 'Mum's family is quite interesting after all, but it's a pity that they didn't hang on to the Book of Kells!' Buoyed up by this unexpected piece of genealogical coincidence, I returned home, turned on the computer and began to type.

Like my daughter, so many others have pondered the age of living organisms and also of the Earth. Biologists can examine the ontogeny of an organism for an indication of its age. As growth proceeds, the individual or colonial organism undergoes change. We are all aware of the stark changes in humans that distinguish infants from pre-pubescent children, and adolescents from fully grown adults. With adulthood these changes become less perceptible, but occur nevertheless. Hair colour changes, hair loss in many males increases, ears in men often become larger, and so on. In humans, it is easy to determine the age of an individual simply by asking, although this may still draw a blank. It is perhaps somewhat indelicate to ask the elderly their age. If they refuse to answer, or worse still cannot remember, one can raid the desk bureau and pull out the folded and faded birth certificate that will supply the answer. Although similar certificates might supply the information on the age of thoroughbred horses or of Cruft's champions, such certificates do not exist for most of the living organisms on Earth, nor for the inanimate Earth itself. For these we have to rely on other chronological indicators.

The early twentieth-century English microvertebrate palaeontologist W. C. Swinton was interested in Eocene fish, and conducted a careful study of the bones found in their ears. These otoliths are the shape of dinner plates, but much smaller. What he found was that they appeared to be composed of skeleton deposited in concentrically arranged patterns. He showed that these rings could be used to accurately age a fish. Similarly, the horsemen of the Tashkent plains or the wet fields around Ballinasloe in the west of Ireland can tell the age of a prospective purchase by looking into the horse's mouth and examining the condition of its teeth. They can rapidly tell if a horse claimed to be a three-year-old is rather longer in the tooth than that, and consequently worth much less. The age of trees is widely determined by ring counting, and this science of dendrochronology has proved to be a valuable resource in the study of past climates and an indicator of possible future climate changes.

But the Earth has no ears containing otoliths, nor does it have teeth or annual rings. It presents a complex array of indicators which philosophers, scientists and men of the cloth over at least two millennia have examined to answer the question: how old is the Earth?

This book presents the fascinating story of our attempts to determine the age of the Earth on which we all live. Since earliest times we have attempted to understand the nature of the Earth and its formation. Estimates of its antiquity have varied considerably from low biblically derived timescales to recently derived higher ages based on meteorites. Many novel methods have been pressed into service. Researchers have examined the biblical chronologies, the cooling rate of the Earth, rates of erosion and the thickness of sedimentary rocks, the saltiness of the oceans, the radioactivity of the rocks, and the constituents of the Moon and meteorites. All have been important steps in the evolution of this theme, and have contributed to our present understanding of the Earth.

The debate that has been going on for over two thousand years has pitted various protagonists against each other: biblical versus non-biblical chronologers; physicists versus geologists; and more recently scientists versus creationists. At the turn of this present century a consensus has been reached amongst the scientific community and the majority of the general public that the Earth is four and a half thousand million years old.

Can we style these geological and biblical investigators 'chronologers' as I have done in the title of this book? Yes, I believe that it is perfectly acceptable to do so. According to the *Oxford English Dictionary* a chronologer is one versed in chronology; 'One who studies chronology, one who investigates the date and order in time of events' – in this case, the date of the origin of the Earth.

This book examines a number of episodes in the debate, starting with the ideas of some ancient civilisations and finishing with the present state of our understanding of this concept. It does not set out to produce new research facts; rather it brings together the strands of diverse research in geology, astronomy and religious chronology and aims to make the whole story of the dating of the Earth available to a new body of readers not conversant with the scientific literature.

Acknowledgements

I owe a great debt of gratitude to two Fellows Emeriti of Trinity College Dublin, both of whom taught me during my undergraduate years, and both of whom became colleagues once I joined the staff of the college. Gordon Herries Davies is a historian of geology and geomorphology whose writings and lectures captivated and inspired me to embark on studies in his field. He gave me early guidance and huge encouragement when I dipped my toe into the subject and later nominated me for membership of INHIGEO (the International Commission on the History of Geological Sciences). Through this group I have made many friends throughout the world. A number of years ago I was delighted to host a group of INHIGEO colleagues on an excursion around Ireland when we examined those sites of significance to historians of geology such as the Giant's Causeway, and on the second last day, Gordon, together with Jean Archer, interpreted for us the unusual features of the Blackwater Valley. Charles Hepworth Holland was both my teacher and my boss. A stratigrapher and cephalopodologist who focuses on fossil nautiloids, he instilled in me a love of palaeontology and systematic order. He agreed to supervise my doctoral thesis, and despite my efforts he still finds the taxonomy of Carboniferous bryozoans rather perplexing. In truth I cannot claim to understand the complexity of nautiloid taxonomy! He has a wonderful way of encouraging independent research, and allowed me to follow my own rather varied research interests in palaeontology and in history of geology. Unfortunately, in the modern arena where research exercises have assumed too great an importance, many university academics are forced to carry out research in an area which appears to be of greater value to their department in gaining credit than the field to which their instincts take them. I am happy to count both Charles and Gordon as friends, and appreciate all that they have done for me.

Many colleagues, friends and family have helped me in various ways, both directly and indirectly, in the preparation of this book and if I have omitted anyone from this list I sincerely apologise. For their assistance I thank Bill Ausich, Peter Bowler, David Branagan, Bill Brice, Stephen Brush, Duncan Burns, Norman E. Butcher, Albert Carozzi, Stephen Coonan, Gordon Craig, Brent Dalrymple, Bill Davis, Dennis Dean, Hal Dixon, Ellen Tan Drake, Silvia Figueirôa, John Fuller, Greg Good, Gordon Herries Davies, Kathleen Histon, Charles Holland, Mary Spencer Jones, Ted Nield, Marcus Key, Jr, Martina Kölbl-Ebert, Gary Lane, Cherry Lewis (who encouraged me to write this book), Mark McCartney, Donald McIntyre, Ursula Marvin, Eugenij Milanovsky, Nigel Monaghan, David Murnaghan, Sally Newcomb, Chris Nicholas, John Nudds, David Oldroyd, Matthew Parkes, Michael Roberts, Ed Rogers, Martin Rudwick, Ian Sanders, the late Bill Sarjeant, Len Scott, George Sevastopulo, Crosbie Smith, Adrian Somerfield, Margery and Larry Stapleton, Ken Taylor, the late John Thackray, Hugh Torrens, Ezio Vaccari, Gerry Wasserburg, Denis Weaire, the late David Webb, Leonard Wilson, Peter Wyllie, John Wyse Jackson, Michael Wyse Jackson, Peter Wyse Jackson and Ellis Yochelson.

I am also most grateful to those who have supplied images for use in this book, in particular Hugh Torrens and Mary Spencer Jones. I thank the California Institute of Technology for permission to use information from the 1995 interviews conducted with Clair Patterson. These now form part of the Caltech Archives. Every effort has been made to trace copyright holders of images where appropriate.

I am grateful to Matt Lloyd, my commissioning editor at Cambridge University Press, for his encouragement, dedication and patience.

Finally I thank my family, Vanessa, Susanna and Katie, who at times didn't see much of me when I ascended the stairs to the study to write this book, and when they did see me had to put up with discussions about Ussher, Buffon, Kelvin *et al.* I hope they feel that it has been worth the wait.

I The ancients: early chronologies

Creation, be it of the Universe or the Earth, has been a subject of fascination for centuries. Through the ages, philosophers and latterly scientists have struggled to come up with a logical explanation of how the Earth and the Universe came to be. Allied to this has been the question: when did creation take place?

In many cases early philosophers and thinkers made no distinction between the date of formation of the Earth, the Universe or indeed the appearance of mankind. In many mythologies no actual dates are given. Creation myths, or more correctly beliefs, as one would expect, are frequently closely related to the experiences exerted on the civilisations that propounded them. Thus among peoples of the northern hemisphere great emphasis is placed on ice, frost and cold climatic conditions, whereas the Persians and Egyptians set great store, respectively, by the Tigris and Euphrates, and by the Nile, and their essential life-giving properties. These beliefs allowed man to grasp an understanding of his environment and the planet on which he lived. The annual, seasonal, diurnal cycles were seen to be recurring, and these events were explained through the adoption of higher life-forces or gods.

In some civilisations the Earth and Universe are seen as everlasting, while in others they have a definite time-progression from birth to eventual death. Nearly 2,000 years ago the Roman poet, writer and philosopher Carus Titus Lucretius (c. 95–55 BC) published *De rerum natura* just two years before his suicide. In this important poem he made several observations about the Earth and natural history, including suggesting that clouds formed from moisture, that volcanoes developed as winds inside the Earth heated up rock and produced magma, and that earthquakes were also triggered by these internal winds. He also pondered the planet's history, saying: 'the question troubles the mind with doubts, whether there was ever a birth-time of the world and whether likewise there is to be any end.'

Creation and the processes by which it happened were often explained through the incarnation of deities. The Egyptians had a whole pantheon, paralleled to some degree by the Greek and Roman gods. Even the Celts had their own line-up of gods, many of whom were related to the natural elements and astronomical bodies. Various peoples used these ideas to rationalise their existence – to understand their position within the environment, and the various elements (air, land and water) that constituted that environment. They also used beliefs to derive a cosmology or history of their planet that they themselves could understand.

Creation and the early history of the Earth have been the subject of mythological stories derived from many cultures. Certainly these ideas would have developed independently of each other. Today when we refer to 'myths' the general understanding is that these were ideas that are now discredited or wholly incorrect. A search on the Internet under 'creation myths' certainly leaves this impression. Here I prefer to use the term 'beliefs' instead of 'myths', reflecting the older but now largely superseded concept of the latter term. There is no doubt that the beliefs outlined below were of huge significance to the various civilisations in which they evolved. There is no evidence to suggest that these peoples considered these ideas fallacies. While modern scientists are confident that our understanding of the Earth's creation and its progression are broadly understood and explained in a logical manner, there is of course a possibility that we, like our predecessors, are incorrect. I, for one, believe that the Earth has a very long history and that geologists and astronomers have got the story correct. Others, perhaps, do not feel as confident.

EGYPTIAN BELIEFS

The oldest documented creation beliefs are those of the Egyptians, and can be traced back to around 2,700 BC. There are several strands or traditions and they have become somewhat interwoven, but all have a common thread in that the creation schemes proceeded in stages. Those stories from the cities of Heliopolis, Hermopolis and Memphis are the most important. Heliopolis lay north of Cairo on the confluence of a major divide of the Nile as it begins to widen into its delta, and its population was held in the grip of a Sun cult. At the beginning, Nun, the god of the primordial waters and father of the gods, caused a mound of dry land to emerge from the primordial chaotic water. On the land stood Atum, who created himself, and then the twins, Telfnut the goddess of moisture, and Shu the god of air, who became the parents of Geb the god of the Earth and his sister Nut the goddess of the Sky. When Shu discovered that the siblings had secretly married, he became angry and with great force separated them. With the assistance of two ram-headed gods, Shu raised Nut into the sky, and subjugated Geb beneath his feet, where he lay with his limbs bent – these symbolised the mountainous undulations of the Earth's crust. Atum was later considered to be the god of the setting Sun, and Ra, one of the most important of all Egyptian gods, to be the god of the risen Sun.

From Hermopolis, a city south of Cairo on the western bank of the Nile now called Matarea, came two creation stories. The first starts, like that of Heliopolis, with the emergence of land from chaotic waters. But it then tells of the appearance of an egg that hatched and yielded the Sun whose rise into the heavens was followed by the creation of all living matter. The second tradition saw the replacement of the egg with a lotus bud that floated on the surface of the waters. Horus the Sun god emerged from the opened petals of the lotus, and his rays radiated throughout the world. The story from Memphis, which is just southwest of Cairo on the left bank of the Nile, is rather different, and simpler than those from Heliopolis and Hermopolis. Creation was effected by the creator god Ptah (Figure 1.1) who in his heart thought up the concept, and having spoken of it brought the Earth into being.

CHALDEAN AND BABYLONIAN BELIEFS

Chaldea was the ancient name for the area of what is now southern Iraq, an area enclosed by the great rivers Tigris and Euphrates

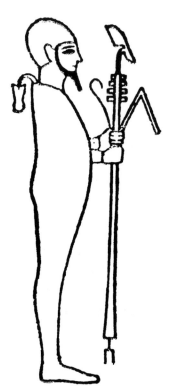

Figure 1.1 Ptah, the Creator God from the mythology of Memphis. He is shown holding a sceptre the head of which combines the *was* and *djet-pillar* symbols – the former had a forked base and was topped with the head of a dog, while the latter possibly represented a tree from which the leaves had fallen. From Ptah's neck a menat hangs down his back (from Anon., *Helps to the Study of the Bible.* (Oxford University Press, 1896), Plate 25).

northwest of their confluence, before they empty into the Persian Gulf. Later it was incorporated into a slightly wider region that became known as Babylonia; the term 'Chaldeans' in the Old Testament was often applied to astrologers and astronomers, and in general elsewhere to magicians. The notion that the Universe, and by inference also the Earth, had a cyclical history, originated in Chaldea. Each cycle was known as a Great Year (although it was certainly longer than a year as we understand it to be) which began and ended in either flood or fire. The later Babylonian myth of creation was encapsulated in the *Epic of Creation* inscribed in cuneiform lettering on six tablets that were found in the ruins of the Library of Assur-bani-pal (668–626 BC), King of Assyria, in the city of Nineveh (Figure 1.2). A seventh tablet was added in AD 142. The Epic recalls the actions of the god Marduk who was the only god capable of defeating Tiamat, the

Figure 1.2   Tablet with cuneiform inscriptions from the library of Assur-bani-pal, King of Assyria, telling part of the creation story of the Universe (from Anon., *Helps to the Study of the Bible*, Plate 57).

dragon of Chaos. In the beginning the god Apsu and Tiamat came together and bore the gods of Earth and Heaven. These offspring attempted to bring some order to their parents' chaotic lives, but conflict followed and numerous deities were killed and replaced with others. Marduk, who was the son of Ea, the god of water, armed himself with thunderbolts and lightning and, with the assistance of the winds, went into battle against the eleven monsters created by Tiamat who were under the command of her husband Kingu. Eventually Marduk prevailed, killing the dragon and dividing her body into two. One half became the heavens while the other became the Earth and the oceans. Plants and animals were then created, and followed by Man who was formed by Ea from clay and the blood of the god Kingu. It is not clear when creation occurred, but man, according to the Babylonians, appeared half a million years ago.

INDIAN OR VEDIC CREATION BELIEFS

Vedic faiths are those that arose on the Indian subcontinent, the oldest of which is Hinduism, followed by the later Buddhism and Jainism. Essentially all three faiths regard the Universe as having no beginning nor end.

In Hindu belief the Universe developed from the Hiranyagarbha or golden egg, which brought into existence the supreme god, the Brahman ('spirit' in Sanskrit). The egg contained the continents, oceans, mountains, the planets, the Universe and humanity itself. After a thousand years, the egg was said to have opened, releasing

Brahman, who began the work of creation. He found that the Earth was submerged under primordial waters, so he dived into the waters, and having assumed the form of a wild boar, he used his tusks and dragged the land up so that it lay above the surface of the water.

Time in the Universe comprises a series of ever-repeating cycles from birth, to growth, decline and death, followed by rebirth and the commencement of a new cycle. But how long is this cycle? Certainly millions of years. In order to indicate this immensity of time to the general populace, the storytellers told of a man who once every hundred years went to the top of the mountain and rubbed it with a cloth. The time that he would take to wear the complete mountain away was shorter than one universal cycle from birth to death.

Hindu duration of the Universe

In Hindu tradition the beginning of each cycle is announced by Shiva the Lord of the Dance, who bangs a drum held in his right hand. The ageing cycle ends in the flames held in his left hand, when all is absorbed into Brahma, and a new cycle commences. Each of the four ages of the world is called a Yuga and the four combined are termed Mahayuga or 'Great' Yuga. Each cosmic cycle comprises one day and night in the life of Brahma. The day lasts a Kalpa or 4,320,000,000 years and the night an equivalent time. In a Kalpa there are fourteen periods called Manvantaras each presided over by a special cosmic deity. The lifespan of Brahma is thought to be 100,000 daily cycles, and so to Hindus, the Universe and Earth are many billions of years old.

Buddhist beliefs

Buddhism was founded in the sixth century BC in northeast India by Siddhartha Gautama (563–483 BC) who was given the title 'Buddha'. Although Buddhists believe that the cosmic cycles continue unabated, there is possible release from them if 'Nirvana', a state of happiness or peace, is reached.

CHINESE AND JAPANESE BELIEFS

Some Chinese philosophers argued that Earth history was cyclical, and that each cycle took 24 million years to complete. It is not clear, however, how many cycles were involved. According to Chinese legend the first man on Earth was called P'an-ku. Later various parts of his body became mountains at the cardinal points of the compass; his arms became the north and south mountains, head was at the east, while the mountains of the west were formed from his feet. His eyes became the Sun and the Moon, and mankind developed from vermin that covered his body.

Early Chinese thinkers considered the Earth was square, some 233,575 steps in length and width. Later, in about AD 723, the mathematician I-Hsing measured its diameter and clearly understood that the Earth was a sphere.

Japanese mythology tells that at the creation of the Earth and the sky, three gods were self-formed, but immediately hid themselves from view. The young Earth, which had a jelly-like consistency, floated on water and from it grew a plant similar to a bullrush which produced two further gods. These, like their earlier counterparts, hid themselves. Following this a series of gods emerged in several generations, and the last two, Izanagi and Izanami, were given the job of consolidating the mobile Earth, and ensuring that its soil was suitable to grow crops. They took a stick and stirred the waters. When the stick was withdrawn, a drop of congealed matter fell back into the water and formed the island of Onokoro, where the two gods made their home. They became attracted to each other but before they could form new islands they had a disagreement because Izanami had spoken first, and being female she should not have done so. Nevertheless they had a child and this became the island of Awa. The couple asked the gods to mediate in their dispute, and following reconciliation they had more children who became either further islands that now make up Japan, or more gods, such as those of wind, the mountains and trees. Their last child became the god of fire. His birth was difficult and resulted in the death of his mother. Izanagi was livid and beheaded the child, whose blood became eight more gods.

GREEK BELIEFS

Greek culture and thinking has a long history that stretches back to the sixth century BC. The earliest writings about the Earth and its chronology were by Hesiodus, the father of Greek didactic poetry, who was born at Ascra near to Mount Helicon in about 850 BC. After an early career as a farmhand he began to write poetry, having been commissioned to do so by the muses. Following the death of his father he fell out with his brother and emigrated. His most famous works are *The Works and the Days*, a poem with an agricultural theme running through it, which contains a section entitled 'Five ages of the world', and *Theogony*, in the first portion of which he describes the emergence of Earth (Gaea) from Chaos. Hesiodus' life ended violently with his murder and his body was thrown into the sea, only to be returned to the shore by dolphins. His dogs found the murderers of their master and threw the two guilty men into the sea where they drowned.

Later thoughts on the chronology of the Earth can be attributed to the Ionian natural philosopher Anaximander (610–547 BC). He was born in the town of Miletus which is situated south of Ephesus in what is now Turkey. Apart from writing about the nature of time and the Universe, and introducing the sundial into Greece, he devised a system of cartography and so is styled by some commentators as the 'inventor of maps'. Anaximander considered that time was endless, but that the Earth's history was cyclical – it and the Universe were being continually destroyed and subsequently reborn. The Universe and Earth were derived from an endless mass of matter, from which evolved a ring of fire comprising the stars, Sun and Moon that enclosed the Earth in its centre. Anaximander was perhaps the first commentator on evolution, nearly two and a half millennia before Charles Darwin. He said that all terrestrial animals had arisen from amphibians, but that humans had evolved from fish. It was natural, in an area prone to earthquakes, that the early Greek philosophers should have an opinion about the structure of the Earth, although there was some confusion and difference of opinion as to where matter had come from. Heraclitus (540–475 BC), a philosopher of Ephesus, who was happy in his own company and shunned others, and who built his home on a dung-heap which provided underfloor heating, argued that all matter originated in fire and that it was never destroyed but simply reorganised. His contemporary Anaximenes (d. 504 BC) held that matter originated in air, and he believed that the Earth was a flat disc around which rotated the stars and planets. Later still Xenophanes (560–478 BC) regarded fossils as being proof that land had once been submerged, and Empedocles (490–430 BC) said that the Earth had developed in stages, that its core was molten – a fact not confirmed until the middle of the nineteenth century – and that the Earth and all on it was constantly in a state of change. These fifth-century BC philosophers together with Pythagoras (580–500 BC), best remembered for his laws of trigonometry, resurrected the theory of a cyclical Great Year that had been formulated by the Chaldeans. Herodotus (484–408 BC) is best known for his nine-volume history of the known world. However, he also made some geological observations and was aware that land was produced by sedimentation, and calculated that it would take 5,000 years for the Red Sea to silt up completely.

SCANDINAVIAN BELIEFS

In the northern latitudes of Scandinavia and Iceland, creation beliefs drew on the physical characteristics of the land. Initally there was no Earth, nothing but a large abyss. The first worldly place was a land of mists and clouds called Niflheim which was situated in the north, and in which spurted a great fountain that was the source of the Twelve Rivers. These carried very cold water towards the south, where Muspellsheim, the land of fire, was situated. Through this land flowed rivers in which a strange material slowly hardened and set. When it came into contact with the northern rivers a frost covered this material, and slowly the frost began to fill the abyss. However, warm southerly winds caused some of the ice to melt and from the meltwaters Ymir, the first human, formed. When he was asleep two more giants formed and these were fed from the milk of Audumla, a cow also metamorphosed from the meltwaters. The offspring of the giants included Odin who rebelled against Ymir and killed him. His body became the landmass known as the middle Earth. One can clearly see the geological influences on this story. The cold northern rivers are most probably the cold Arctic current that when it reaches the warmer waters of the Atlantic produces thick fog banks, particularly around Newfoundland. The warm rivers with the solidifying matter are most probably lava flows which would have been known about from Iceland.

EARLY CREATION BELIEFS FROM THE AMERICAS

The Mayans, who occupied the northern portion of the Central American Peninsula area around Guatemala and southern Mexico, considered that the Universe was cyclical and that it could be destroyed and recreated. It formed initially from an ancient sea. Later the gods of the Sea and Earth who occupied this early ocean were joined by the gods of Newborn Thunderbolt, Sudden Thunderbolt and Hurricane and they decided to create land from the waters, after which the Sun, Moon and stars were formed. Difficulties with man soon occurred, and the gods attempted at least another four times to perfect Creation. The date of the last Creation has been given as 5 February 3112 BC.

CREATION BELIEFS FROM THE PACIFIC

Various island groups in the Pacific have their own individualistic creation stories, and frequently more than one story is associated with each island. Naturally enough given the strong geographical influences on the islands, many of the stories have a common thread. In Polynesia the Earth was born out of surrounding water. Maui, a major folkhero, reeled up New Zealand, Hawaii and Tonga. Some of the island chains were produced by his fishing, at different times, or when (as in the case of Hawaii) one large piece of land broke up as it was hooked out of the water on his fishing line. In the Admiralty Islands a snake that floated across the surface of the water was thought to have produced its island chain.

Some cultures believe that their islands were dropped into the ocean from above. The Hawaiian view is that Hawaii formed after a bird dropped an egg into the Pacific. The wood shavings discarded by a god as he laboured in his heavenly workshops gave rise to the island of Tonga, while sand scattered on the ocean produced several island groups including parts of Sumatra.

Geographical characteristics of certain islands were also attributed to folklore. In Borneo the valleys were said by the Kayan people to have been excavated by a giant crab using his pincers rather like a hydraulic grab on a modern mechanical earth-mover; the crab was thought to have fallen from the sky rather than come out of the sea. Other ideas from Borneo explained the unevenness of the Earth by involving two birds and two eggs. One egg became the heavens while the other became the Earth. Unfortunately the latter was larger than the former which was supposed to surround it. Unperturbed, the birds crushed the Earth egg so that it could be enveloped by the heavenly egg, and this crushing produced the mountains and valleys familiar today.

Many of these stories are very ancient, and most follow a similar pattern: they evoke a higher being or god. As early thinkers did not travel widely they had little perception of the vastness and complexity of the regions in which they lived and formulated these ideas. The ability to think beyond the human condition is an advance in terms of philosophy, but frequently it was found that many natural features could not be logically explained. It was difficult enough thinking about local topography without having to take on a global perspective. The development of creation stories circumvented the difficulties of having to understand the origin of the Universe and the Earth. The invocation of gods and deities allowed the inexplicable to be reasoned without having to delve too deeply.

This style of creation belief was not forcibly challenged until the seventeenth century when the science of geology began to emerge as a separate scientific discipline. Then, and in the two centuries that followed, many deep-seated religious and cultural beliefs were examined and scientifically challenged.

2 Biblical calculations

In any second-hand bookstore one can find old copies of the King James Bible, often bound in black leather with gilt-edged pages, shelved high up out of reach. These neglected volumes were more often than not prizes awarded by some religious group such as the Attercliff Baptist Sunday School for proficiency in answering questions on scripture, and signed by the local cleric and superintendent. How many of the beneficiaries of such prizes would have noticed the odd inscription 'Before CHRIST 4004' or '4004 BC' printed in black or occasionally red ink either in the margins or between two columns of versified text at the opening of the Book of Genesis? If they did, what did they make of it? What questions did this figure conjure up in their minds and how often was it debated on Sundays?

Shout '4004 BC' in a crowded lift travelling to the fifteenth floor of an office building, or in a coffee shop on a Sunday morning, and 50% of the occupants will think they are sharing space with some crazed individual and want to get as far away from you as possible, while the other 50% will probably think either 'Creation' or 'Ussher'. As many people know, 4004 BC is the date of the Creation arrived at by James Ussher (1580–1656), Archbishop of Armagh, in Ireland. However, he was not the first person to attempt to date Creation using the pages of the Bible, nor was he the last.

One of the first biblical computations was by Theophilus of Antioch (d. 191) who converted to Christianity in adulthood and was later consecrated Bishop of Antioch, in present-day Turkey. He wrote an important tract *Ad Autolycum* (To Autolycus) in which he stated that the age of the Earth from its creation to the time that he wrote his letter was just over 5,698 years. The longest duration was 2,242 years, which represented the period from the Creation to the Flood.

ELIZABETHAN AND SEVENTEENTH-CENTURY
BIBLICAL CHRONOLOGIES

In the 1600s many commentators were constrained by their readings and interpretation of the first chapter of Genesis, which in the King James version began: 'In the beginning God created the heaven and the earth', and ended 'And the evening and the morning were the sixth day.' Soon, nonetheless, it was abundantly clear that to take such words absolutely literally did not make sense, and the biblical duration of one day was generally taken to represent an actuality of 1,000 years. As a result, a severe constraint was placed on the duration of time since the Creation. It was also recognised by contemporary thinkers that the creation of the stars and extraterrestrial matter, or the Universe, took place earlier than that of the Earth by some longer undefined time period, although this is not obvious to readers of Genesis. Most commentators took up an anti-Aristotelian stance. They also had to grapple with the vagaries and variations of chronology that the various versions of the Bible presented: in the Hebrew Massoretic text, the period between the Creation and the Flood lasted 1,656 years and Abraham appeared 2,083 years after Creation whereas in the Greek text, the Septuagint, the Flood washed the surface of the Earth 2,262 years after its formation and Abraham was born in 3,549. In the position of Abraham in history this is a difference of 1,466 years. The Septuagint version dates back to the second century BC and was used by Jews who had emigrated to Alexandria. It is still used in the Greek Orthodox Church. The Massoretic version was adopted by Hebrew scholars 200 years after Christ's death, and its versions of the Old Testament were largely incorporated into the King James or Authorised Bible of 1611. The third version of the scriptures was the Latin Bible or Vulgate, which was the version translated by Jerome in the fourth century AD. This version was largely used in Elizabethan England.

The complex mathematical and textual difficulties facing these chronologers were aptly noted by the cleric Thomas Allen (1608–1673) in the preface to his 1659 book *A Chain of Scripture Chronology*: 'The

World, which his hand made, is aged; but of what age, who can justly tell?' William Nisbit, a Scottish cleric, in 1655 remarked in *A Scripture Chronology*, 'There is great disagreement among chronologues in counting the years from the Creation of the World to the death of our Saviour.' However difficult these calculations were to reconcile, Thomas Allen was sure that the answer lay in examining the biblical texts: 'The *Sacred Writ* is the best *Register*: Therein its *Age* possibly may be found; but so *various and discrepant* are the *Calculations* of men, that it may be ranked amongst *unsearchable*', but was not confident that even the highest intellectuals could derive from it a correct date for the Creation.

In England one of the earliest attempts to estimate the time since the Creation appeared in John Swan's *Speculum mundi* in 1635, which combined a description of his natural surroundings with a timescale into which they were placed. Swan, who was rector of a small parish near Cambridge, put a duration on the six segments of biblical events first outlined by Robert Grosseteste in his *Hexaemeron* of 1225. The first five ages of the six were derived from the Old Testament and Swan calculated that these had a duration of around 3,997 years. Since the crucifixion of Christ 1,635 years had elapsed, and since the Creation 5,632 years. Swan's placing of this chronology within a framework of the nature of living organisms and the slow decline through time sets it apart from the purely biblical chronologies that followed.

A COMMONWEALTH OF CHRONOLOGIES

England in the 1640s was in the grip of Civil War. With the defeat of the Royalist army on the field at Naseby on 14 June 1645 the Civil War was moving towards its end and Charles I surrendered to the Scottish army. But by January of the following year he had been handed over to the Parliamentary Commissioners at Newcastle, and it was only a matter of time before he met his final fate. His trial, held in January 1649, was swift, lasting only seven days – a rather short time by modern standards for a celebrity trial – and he was executed on the 30th of

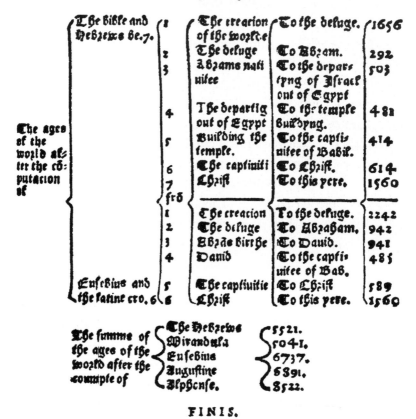

Figure 2.1 *The computacion of the ages of the worlde* published
in *Cooper's Chronology* in 1560. Courtesy of David Branagan, Sydney.

the month. The Commonwealth, which had replaced the abolished
monarchy, lasted eleven years and was a period of Puritanical rule
headed by the Lord Protector, Oliver Cromwell (1599–1658).

At about this time chronologies based on the sacred texts frequently started to appear in print although earlier chronologies were
known, such as that published in *Cooper's Chronology* in 1560
(Figure 2.1). Certainly the chronology published by Archbishop
James Ussher (1580–1656) (Figure 2.2) is the best known. A search on

Figure 2.2 James Ussher (1580–1656) (from J. A. Carr, *The Life and Times of Archbishop Ussher* (1895), frontispiece).

the Internet in the Google search engine using the tag 'James Ussher' or the variant 'James Usher' throws up 23,700 and 4,260 hits respectively, many of which discuss his chronology. Consequently one might imagine that his was one of the few seventeenth-century estimates of the Earth's antiquity. This presumption is completely erroneous: there were numerous chronologies besides Ussher's. Nesbit wrote that 4000 BC was near to the truth (the true date of Creation) and he documented the findings of earlier authors: Philip Bergomensis, 5232 BC; Vernerus in the *Fasciculus temporum*, 5233 BC; Mr Pont, the Englishman, 3981 BC; Heinrich Bullinger (1504–1575), who had introduced Protestantism to Zurich, 3004 BC; Rodolphus Gualterus (1519–1586), the Swiss theologian, 4002 BC; Benito Arias Montano (1527–1598), the Spanish biblical scholar, 4084 BC; Andrew Willet in his hexapla on Chapter 9 of the Book of Daniel, 3967 BC; and Joseph Justus Scaliger (1540–1609), 3985 BC. Hugh Broughton (1549–1612), a feisty and difficult scholar from Shropshire, published *A Concent of Scripture* in 1588 in which he stated that the biblical chronology was a true record and that the Creation happened in 3960 BC.

Table 2.1 *Thomas Allen's 1659 chronology from Creation to the death of Christ.*

Period	Dates	Duration
(1) Creation to the Flood	AM 1–1656	1,656 years
(2) Flood to the Promise to Abraham in Ur	AM 1656–2078	422 years
(3) Promise to the Law (Ten Commandments)	AM 2078–2508	430 years
(4) Law to the building of Solomon's Temple	AM 2508–2988	480 years
(5) Temple to the captivity in Babylon	AM 2988–3276	288 years
(6) Captivity to the Return	AM 3276–3478	202 years
(7) Return to the death of Christ	AM 3478–3968	490 years

AM = *Anno Mundi* or 'The Year of the World'

Roger Drake (1608–1669), a medic who changed career and became a minister, published a *Sacred Chronologie* in 1648 which announced on the title-page that the vast body of time 'From the Creation of the WORLD, to the Passion of our Blessed SAVIOUR' was almost four thousand years. Thomas Allen, whom we have already met, produced a 240-page chronology, which like many others of that time was divided into seven periods and he tabulated the duration of these periods as shown in Table 2.1.

As we can see Christ died in AM 3968 according to Allen, who additionally notes that Christ was born in AM 3934. Creation took place in 3934 BC. Nesbit had produced a similar chronology four years earlier than Allen, again with seven periods, and with an identical time-span for the first period, but after the Flood the durations of the various remaining periods are at variance with each other. Approximately 4000 BC was the generally accepted date for Creation in the mid 1600s. Even on throw-away publications such as the *Dove speculum anni*, an annual almanac produced by Jonathan Dove in London in the 1670s, we find acceptance of this timespan; printed on the cover of the 1677 booklet is '5681', the number of years that had elapsed since the Creation. Many people would have therefore been aware of the figure 4004; not just those who could afford the expensive books, but the many citizens of London who wished to know the dates of the feast days, the phases of the Moon and the hours of daylight in 1677, or who were simply curious to know the signs of the Zodiac or learn some recent history. They would have learnt that in 1646 'The treacherous Scots sold the King to the Parliament for 400,000 pounds'. It is unlikely that Dove's almanac was distributed north of the River Tweed and the Solway Firth.

However, not all chronologies were acceptable to scholars. *Unica vera et infallibilis chronologia biblica* published in 1670 by the Berlin scholar and oriental linguist Christian Ravis (1613–1677) was mocked by all theological scholars who bothered to read it. Ravis was widely travelled: having solicited support for travels in the east, including the offer of a stipend of £24 granted by Ussher in return for locating various manuscripts that the cleric wished to read, Ravis ended up in modern-day Turkey. He returned to London with about 300 manuscripts, some (though not all) for Ussher, to seek a job as a language teacher and made an unsuccessful attempt to gain the Chair of Arabic at Oxford, although he was later to hold similar chairs in Uppsala and Frankfurt where he died in 1677.

THE ARCHBISHOP FROM DUBLIN

Watching Charles I at his execution, from the roof of the Charing Cross house of the Dowager Countess of Peterborough, was a cleric who had been in enforced exile from his native land. This was the aged and now frail Ussher, the Archbishop of Armagh, who having served as chaplain to the King found the whole episode highly distressing and would have collapsed but for the intervention of his servant and chaplain. What was the cleric doing in London? He had had to flee Ireland in 1641, but through considerable diplomatic skill managed to stay on the right side of the royalists and subsequently Oliver Cromwell.

The future archbishop was born at 57 High Street, Dublin, on 4 January 1580 to Aarland and Margaret (née Stanyhurst) Ussher, members of a powerful merchant family. His family name is still well known in the city and has given its name to several streets – Usher Lane, and Usher Street, originally known as Dog and Duck Yard – and to Ussher's Island, one of the south quays of the River Liffey. The latter was the location for *The Dead*, one of James Joyce's most celebrated short stories, published in *Dubliners* in 1904.

At the age of thirteen, James was one of the first students to enter Trinity College, the only constituent college of the University of Dublin, of which his uncle Henry had been named the first Fellow. This university had been founded the previous year by Queen Elizabeth and conveniently found a home to the east of the city at Hoggen Green, in an monastery that had been confiscated by her father. James clearly excelled in university and was appointed a Fellow of the College at the age of twenty (although he resigned five years later), and Professor of Divinity in 1607. Trinity College possesses a wonderful library. Many of the oldest books contained within its limestone walls were purchased by Ussher on behalf of the college on several trips made to England for this purpose. It was appropriate that following his death his own valuable collection of books and manuscripts came to reside permanently in Trinity College. There had been some wrangling about what should become of this collection: the King of Denmark wished to purchase it, but Cromwell intervened and eventually it was purchased for £2,200 by his army stationed in Ireland. Following a period when it was stored at Dublin Castle, it was deposited in the college after the Restoration.

Like most contemporary Fellows of Trinity College Dublin, who were the teaching academics of the day, Ussher took holy orders, and was appointed a vice-chancellor of the university in 1615. For a long time these men had to take a vow of celibacy in order to retain their fellowship, but in a typical Irish twist, the *College Statutes* made provision for their sons to receive their university education free of charge, so long as the boy was registered under his mother's name. Today few Fellows become ordained following their elevation.

By 1620 he had come to the attention of the king, and was appointed Bishop of Meath, a diocese north of Dublin. His rise in the ecclesiastical hierarchy in Ireland was remarkable and he was translated to the archbishopric of Armagh four years later. Ussher's primacy overlapped with a long period of considerable unrest in Ireland. In 1566 the city of Armagh and its cathedral had been sacked and burned by Hugh O'Neill and his followers, so Ussher lived in the town of Drogheda some thirty miles southeast. His daily routine consisted of taking prayers four times and he preached every Sunday. This sermon was later repeated by one of his chaplains to Ussher's servants and any of the local townspeople who cared to attend. When he was not writing and thinking, Ussher was a keen conversationalist, and enjoyed riding and walking. He was married to the daughter of Luke Challoner, one of his university colleagues, and had several children. In 1640 he and his family were visiting England, and the rebellion in Ireland that flared up the following year forced them to remain: Ussher never returned to Ireland, and he spent the remainder of his life in a peripatetic existence in England.

The eminence and standing that this cleric had in England is borne out by the fact that on Ussher's death, on 21 March 1656 at Lady Peterborough's house in Reigate, Cromwell himself ordered that he be buried in St Paul's Chapel in Westminster Abbey and granted his family £200 for funeral expenses. The moving and solemn occasion was attended by so many mourners and members of the nobility and gentry that a military guard had to be employed to ensure their safety. As would be expected the funeral was conducted according to the rites of the Church of England, but this was a unique occasion, in that it was the only time during the Commonwealth that this form of service was used at Westminster Abbey. Following the funeral Ussher was laid in an unmarked grave close to his old tutor Sir James Fullerton (d. 1631) who, like the archbishop, had been a Fellow of Trinity College Dublin.

Ussher was a scholar of the highest calibre. He published numerous books on theology and history including *Britannicarum*

ecclesiarum antiquitates (1639) which gave an account of the supposed visit of St Patrick to Rome, and his most famous work *Annales
veteris testamenti*, published in London in 1650 with a second volume
appearing in 1654. It bore the impressive and long title: *Annales
veteris testamenti, a prima mundi origine deducti: una cum rerum
Asiaticarum et Aegyptiacarum chronico, a temporis historici principio usque ad Maccabaicorum initia producto*, and became one of the
most influential and significant books published in the seventeenth
century. Several imprints were printed simultaneously by J. Flesher
for a number of London booksellers including L. Sadler, who had
his premises in the district of Little Britain, G. Bedell of the Middle
Temple, and J. Crook and J. Baker of St Paul's Churchyard, all of whose
names were carried on the bottom of the title page of their respective
imprints. Later Latin editions under the title *Annales veteris et novi
testamenti* ... appeared in various editions from 1658; a Paris edition
appeared in 1673, and last year a handsome copy of the 1722 Geneva
edition was offered for sale priced at over $1,000.

The 1650 book was translated into English in 1658 as *The
Annals of the World. Deduced from the Origin of Time, and contin-
ued to the beginning of the Emperour Vespasians Reign, and the totall
Destruction and Abolition of the Temple and Common-wealth of the
Jews. Containing the Histoire of the Old and New Testament, with
that of the Macchabees. Also all the most Memorable Affairs of Asia
and Egypt, and the Rise of the Empire of the Roman Caesars, under
C. Julius, and Octavianus. Collected From all History, as well as
Sacred, as Prophane, and Methodically digested*, and printed by
E. Tyler for J. Crook and G. Bedell, booksellers. It is simpler to refer
to it as *Annals of the World*.

WHAT DID USSHER ACTUALLY WRITE?

Over the three and a half centuries that have elapsed since the publication of Ussher's chronology he has often been denigrated for his calculations. Some early commentators did not agree with his findings,
but he did find support among other contemporaries including the

English cleric John Milner (1628–1703) who published the book *A Defence of Archbishop Usher* in 1694. Examples of misquotation of Ussher's work abound elsewhere in the literature. William Brice of the University of Pittsburg at Johnstown, a town in central Pennsylvania chiefly remembered for the terrible loss of life in the floods of July 1889, published a wonderful piece of scientific detection in 1982 in the *Journal of Geological Education*, in which he quoted a number of the mistakes in the factual reporting of Ussher's work:

- 'Creation had taken place on the twenty-sixth of October, in the year 4004 BC, at nine o'clock in the morning.'
- 'October 29, 4004 BC'
- 'In 1654 ... Ussher announced with great certainty that ... the world had been created in the year 4004 BC'

We can add to these Stephen Jay Gould's contribution. He published an essay with the derogatory title 'Fall in the House of Ussher' in *Natural History* in November 1991:

- 'Ussher ... had the audacity to name the date and hour: October 23 at midday.'

There was even a mind-boggling error published in the *Irish Times* editions of 22 October 1996, the 6,000th year from Ussher's date of Creation. One of its journalists informed its readers that 'an early tea would be advisable' as the world was going to end that evening at 6 p.m. Ussher never predicted when the Earth would end.

To give Gould credit, he does not castigate Ussher for his calculation but heaps praise on the Archbishop's shoulders for his chronology which is 'within the generous and liberal tradition of humanistic history, not a restrictive document written to impose authority.' Ussher was a man of considerable scholarship and ability, and his chronology was based on many years of study of the biblical texts, Mediterranean texts, calendars and other sources. He arrived at his conclusions following lengthy explanations. To be certain of what Ussher wrote it is necessary to consult the actual volumes that he

published. Having done this, one can see that Ussher's pronouncements on the date of Creation which appear on the first page of his
1650 *Annales* are straightforward:

> In PRINCIPIO creavit DEUS Cœlum & Terram. [*Genes*.I.i] quod
> temporis principium (juxta nostram Chronologiam) incidit in
> noctis illius initium, quæ XXIII. diem Octobris praecessit, in
> anno Periodi Julianæ 710.

which was translated into English in 1658 as:

> In the beginning God created Heaven and Earth, *Gen*. 1. *v*. 1. Which
> beginning of time, according to our chronologie, fell upon the
> entrance of the night preceding the twenty third day of *Octob*, in
> the year of the Julian Calendar, 710.

and Ussher noted that the first full day was Sunday 23 October, and
this was when God created the angels. The year 710 JP is equivalent to
4004 BC. The Julian Period was a timescale of 7,980 years devised in
1583 by Joseph Justus Scaliger, which was based on the Julian Calendar
but from which he had removed the effects of solar and lunar cycles.
The number 7,980 was derived by multiplying 19 (the length of the
solar cycle) by 19 (the length of the lunar or Metonic cycle) and multiplying the product by 15 (the length of the Roman taxation cycle
known as the Indiction). The Julian Day counts days from 1 January
4713 BC, the origin of the Julian Period, and so by 30 June 2005
2,453,551 Julian Days had elapsed. The year 4713 BC was chosen by
Scaliger as the starting point of the Julian Period as this was the last
time that the three cycles began together.

So what Ussher actually said was that Creation took place during the evening of Saturday 22 October 4004 BC, but he did not mention the exact time of day.

In both the 1650 and 1658 versions, Ussher lists dates as mar-
ginal notes on every page, and gives these in three different ways
(Figure 2.3). The left-hand column lists the dates as 'The year of the
World'. These begin at 1 and increase throughout the book, while on

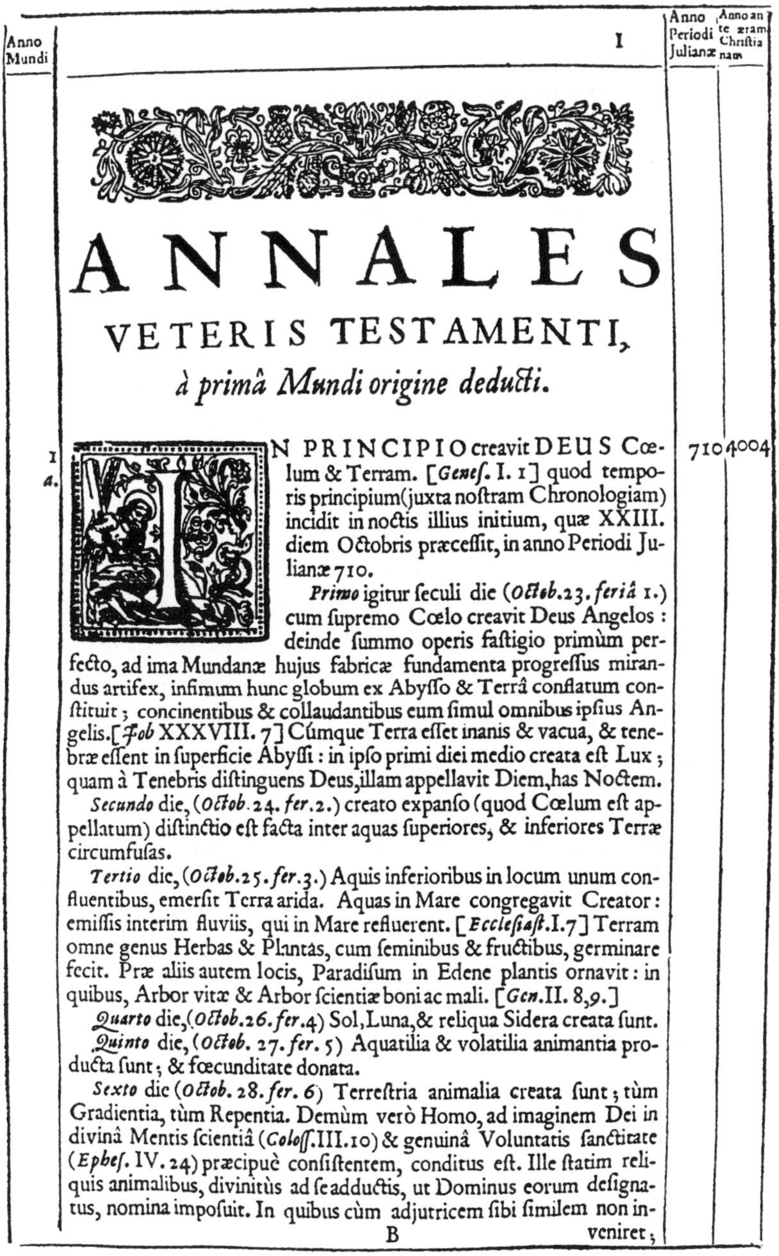

Anno Mundi	I	Anno Periodi Julianæ	Anno ante æram Chriſtianam

ANNALES

VETERIS TESTAMENTI,

à primâ Mundi origine deducti.

1 a. IN PRINCIPIO creavit DEUS Cœlum & Terram. [*Geneſ.* I. 1] quod temporis principium (juxta noſtram Chronologiam) incidit in noctis illius initium, quæ XXIII. diem Octobris præceſſit, in anno Periodi Julianæ 710. 710 4004

Primo igitur ſeculi die (*Octob.*23.*feriâ* 1.) cum ſupremo Cœlo creavit Deus Angelos: deinde ſummo operis faſtigio primùm perfecto, ad ima Mundanæ hujus fabricæ fundamenta progreſſus mirandus artifex, infimum hunc globum ex Abyſſo & Terrâ conflatum conſtituit; concinentibus & collaudantibus eum ſimul omnibus ipſius Angelis. [*Job* XXXVIII. 7] Cúmque Terra eſſet inanis & vacua, & tenebræ eſſent in ſuperficie Abyſſi: in ipſo primi diei medio creata eſt Lux; quam à Tenebris diſtinguens Deus, illam appellavit Diem, has Noctem.

Secundo die, (*Octob.* 24. *fer.* 2.) creato expanſo (quod Cœlum eſt appellatum) diſtinctio eſt facta inter aquas ſuperiores, & inferiores Terræ circumfuſas.

Tertio die, (*Octob.*25.*fer.*3.) Aquis inferioribus in locum unum confluentibus, emerſit Terra arida. Aquas in Mare congregavit Creator: emiſſis interim fluviis, qui in Mare refluerent. [*Eccleſiaſt.*I.7] Terram omne genus Herbas & Plantas, cum ſeminibus & fructibus, germinare fecit. Præ aliis autem locis, Paradiſum in Edene plantis ornavit: in quibus, Arbor vitæ & Arbor ſcientiæ boni ac mali. [*Gen.*II. 8,9.]

Quarto die, (*Octob.*26.*fer.*4) Sol, Luna, & reliqua Sidera creata ſunt.

Quinto die, (*Octob.* 27. *fer.* 5) Aquatilia & volatilia animantia producta ſunt; & fœcunditate donata.

Sexto die (*Octob.* 28. *fer.* 6) Terreſtria animalia creata ſunt; tùm Gradientia, tùm Repentia. Demùm verò Homo, ad imaginem Dei in divinâ Mentis ſcientiâ (*Coloſſ.*III.10) & genuinâ Voluntatis ſanctitate (*Epheſ.* IV. 24) præcipuè conſiſtentem, conditus eſt. Ille ſtatim reliquis animalibus, divinitùs ad ſe adductis, ut Dominus eorum deſignatus, nomina impoſuit. In quibus cùm adjutricem ſibi ſimilem non inveniret;

B

Figure 2.3 First page of James Ussher's *Annales veteris testamenti* (1650) that carries the dates 1 Anno Mundi, the year before Christ 4004, and 710 of the Julian Calendar.

the right-hand side of the pages, two columns tabulate the dates in a different way. The inside right column lists the dates according to the Julian Period and starts with 710 (these increase through the volume) while the outside right column gives 'the year before Christ' and starts with 4004 (and these decrease through the book). In the 1658 English translation the first two dates in the two right-hand columns on page 1 are transposed so that the Julian Period date is given as 4004. It seems this was a printer's error not noticed by the proof reader, but by page 2 subsequent dates appear in their correct columns.

How Ussher arrived at this date is explained by him (quoted from the 1658 translation):

> We find moreover that the year of our fore-fathers, and the years of the ancient Egyptians, and Hebrews were of the same quality with the Julian, consisting of twelve equal moneths, every of them containing 30 dayes … adjoyning to the end of the twelfth moneth, the addition of five dayes, and every fourth year fix. And I have observed by the continued succession of these years, as they are delivered in holy writ, That the end of the great Nebuchadnezars, and the beginning of Evilmerodachs (his sons) reign, fell out in the 3442 year of the World, but by collation of Chaldean History, and the Astronomical Cannon, it fell out in the 186 year of Nabonasar, and, by certain connexion, it must follow in the 562 year before the Christian account, and of the Julian Period, the 4152[.] and from thence *I* gathered the Creation of the World did fall out upon the 710 year of the Julian Period, by placing its beginning in Autumn: but for as much as the first day of the World began with the evening of the first day of the week, *I* have observed that the Sunday, which in the year 710 aforesaid, came nearest the Autumnal Æquinox by Astronomical Tables, (notwithstanding, the stay of the Sun in the dayes of *Joshua*, and the going back of it in the dayes of *Ezekiah*) happened upon the 23 day of the Julian October; from thence concluded, that from the evening preceding, that first day of the Julian year, both the first day of the Creation, and the first motion of time are to be deduced.

> I encline to this opinion, that from the evening ushering in the first day of the World, to that midnight which began the first day of the Christian æra, there was 4003 years, seventy dayes, and six temporarie howers; and that the true Nativity of our saviour was full four years before the beginning of the vulgar Christian æra, as is demonstrable by the time of *Herods* death.

Essentially what he was saying was that the ancient annual calendars were equivalent to that used in his time, and that working with the older chronologies, and the death of King Nebuchadnezzar, he could pinpoint Creation at 4000 BC, and he added four years as a correction for the actual date of birth of Christ. Ussher was conforming to the general acceptance that the Earth was 6,000 years old.

JOHN LIGHTFOOT

From where did the 'nine o'clock' that various later authors cited appear? Who wrote this, and was it in reference to the creation of the Earth or to some other event? Clearly it had nothing to do with Ussher. The citation of this hour appeared in the writings of John Lightfoot (1602–1675) and it was appended by later commentators to the findings of Ussher. Lightfoot was a biblical scholar who had been born into an ecclesiastical family in Stoke-on-Trent. In 1617 he entered Christ's College, Cambridge; following graduation he entered the church, and spent a short, unhappy sojourn in London before procuring the living of a parish near his home town in 1630. Twelve years later he returned to London where he became involved in the Westminster Assembly, a body of 151 clerics (who fell into four factions) and 30 laymen that had been appointed by Parliament to organise the restructuring of the Church of England. Lightfoot sided with the Erastians who felt that state law should have precedence over church law. The body met over a thousand times but never reached a consensus that was acceptable to the authorities in England, although its findings were accepted by the Church of Scotland. At the same time as arguing with fellow clerics in London Lightfoot took up the living of Much Munden near Hereford, and for a while alternated for short periods between the capital and the country. In 1650 Parliament appointed him Master of St Catharine's College, Cambridge. He caught a cold travelling to the cathedral town of Ely and died there on 6 December 1675; he was buried at Much Munden.

In 1642 Lightfoot published a small twenty-page book entitled *A Few, and New Observations, upon the Booke of Genesis. the most of them certaine, the rest probable, all harmelesse, strange, and rarely heard off before.* He dedicated the work to his fellow inhabitants of Staffordshire – what they made of it is anyone's guess – and to his other friends in London. There on page 4 is printed as verse 26: 'Man created by the *Trinity* about the third houre of the day, or nine of the clocke in the morning.' Nowhere in the slim volume did Lightfoot mention the creation of the Earth; how later commentators mixed up this state-ment, which concerned the appearance of mankind, with that of Ussher's pronouncement on the creation of the Earth (which was published eight years later) is hard to understand!

Following the appearance of *Observations*, Lightfoot produced books on Exodus in 1643 and on the Acts of the Apostles two years later. Between 1644 and 1658 he published the four-volume series *The Harmony of the Foure Evangelists: among themselves, and with the Old Testament.* In the 1844 volume of this series, in the first part of the *Prolegom*, he provided a date for Creation: 'from the beginning of time to this fulness of it, hath laid this great, wondrous, and happy occurrence of the birth of the Redeemer in the yeere of the world, three thousand nine hundred twenty eight'. In fact he was more specific and gave the actual day as the September equinox, or 12 September.

NUMEROUS CHRONOLOGICAL ESTIMATES

In 1809 William Hales documented 156 chronological estimates, and in 1861 Leonard Horner the geologist suggested that in fact the num-ber was closer to double that estimate. These showed a range in time since the Creation of 6,500 years maximum to 3,600 years minimum duration. At about the time that 4004 BC was being bandied about,

Zoroaster, or Zerdusht as he was alternatively known, the seventeenth-century Persian philosopher and founder of the Magian religion, said that the Earth was 12,000 years old.

It is hardly surprising that the durations of the Earth according to these European works cluster around 6,000 years, given that many of the calculations were based on the same versions of the Bible or on earlier chronological calculations such as that of Robert Leicester (c. 1266–1327) whose chronology and history of the Jewish people *De compoto Hebreorum aptato ad kalendarium* of 1294 is still in the Bodleian Library in Oxford. Equally there was a suggestion that each of the six days of Creation was equivalent to 1,000 years. Credence for this argument came from Psalm 90 *Domine, refugium*, verse 4: 'For a thousand years in thy sight are but as yesterday: seeing that is past as a watch in the night.' I suspect that an equally important influence on the congruity of time-spans by these authors is their conservative and conforming nature one to another.

Why so many chronologies? This is difficult to determine, but perhaps one reason was that scholars suddenly became preoccupied with trying to determine when the Earth would end, and when they and their fellow men would be subjected to the Last Judgement. An understanding of biblical chronology would help unravel this conundrum.

THE KING JAMES BIBLE, AND THE RISE AND FALL OF USSHER'S REPUTATION

In churches across England in the late 1500s it was usual to find copies of the Bishop's Bible of 1568, which had been adopted as the official version by the Church of England. Books at this time were rare and expensive, but those people who could afford their own personal copy of the scriptures generally used the lighter Geneva Bible which had been published in 1560. However, by the end of the century various theologians and scholars who were dissatisfied with the translations of these two versions had been pressing the king and his advisors to authorise a new English translation. In 1604 King James I of England

and VI of Scotland authorised the publication of a new English translation of the Bible and charged fifty-four of the most eminent scholars in England with the task of providing a translation. No doubt this was a difficult commission given the large number of men appointed. For the next seven years the scholars toiled and eventually in 1611 the King James Bible or Authorised Version appeared. This version of the sacred texts remained the 'official' English version until the 1950s when New English and 'International' translations of the Bible began to appear.

Insertion of a marginal note
The first bible that carried a chronological marginal date was published in Oxford in 1679. The first appearance of the date 4004 BC in the King James Bible was in those copies published in 1701, in which it was placed adjacent to the opening of Genesis. Who arranged for this to happen? The story is illuminated in a paper published in 2005 in the journal *Earth Sciences History* by John Fuller. He tells us that the idea of inserting such marginal dates was thanks to John Fell (1625–1686), a cleric of Oxford and Dean of Christ Church in the city who proposed this measure in 1672. Advice on which chronology to use was received from William Lloyd (1627–1717), a son of the rectory of Tilehurst in Berkshire, and from 1699 the Bishop of Worcester. He was also the pre-eminent chronologer of the time. Lloyd's own biblical chronology which started at 4004 BC was eventually published in 1731 but under the authorship of his nephew and chaplain, Benjamin Marshall. Within this publication, Ussher's work is listed as being one of the sources consulted but he is not acknowledged as being the source of the date 4004 BC. In reality, as we have seen earlier, this date could have been derived from a number of sources. The first bibles that carry this marginal date do not acknowledge Ussher as its source, although many later nineteenth-century printings do.

John Fuller concludes: 'Ussher's fabled date of creation enjoys an unchallengeable history of citation, radiates an immaculate guarantee of truth, yet is nothing more than a print-driven fallacy.'

If we accept this viewpoint (and given Fuller's well-constructed argument it is difficult to do otherwise) then it follows that Ussher was *not* the originator of the 4004 BC date. However, we need to question if Lloyd's 4004 BC was derived from or heavily influenced by Ussher's work or if it was it calculated independently of it. The romantic in me would like to think that the marginal dates in the King James Bible owed much if not all to Ussher, but this may be impossible to prove.

It is easy with hindsight to see how biblical calculations of the age of the Earth were derived, but the now seemingly ridiculous time-spans presented by the seventeenth-century authors should not take away from the scholarship that underpinned their computation. The integrity of authors such as Ussher, Lightfoot and Lloyd is above reproach, and their achievements deserve to be respected as highly significant contributions to the current debate.

3   Models of Aristotelian infinity and sacred theories of the Earth

Scientific thinking during the seventeenth century was very much influenced by religious belief, dogma and the Scriptures. A number of persons, including some of the cloth, who were for the large part highly educated, began to think about the history of the Earth. But however broad-minded they were, their thinking was constrained by their religious beliefs. Some men, even some Jesuits, were prepared to take some risks with their ideas that could have been interpreted as being counter-religious: it was not long since such men would have been burnt at the stake for heresy.

In Europe in the middle of the seventeenth century, ideas on the nature and history of the Earth began to divide and soon two strands developed. One strand originated with the famous French philosopher René Descartes (1596–1650) and included the German Jesuit priest Athanasius Kircher (1602–1680). They circulated ideas that emphasised the mechanical and chemical processes that they thought explained the features seen on the surface of the Earth, and in some cases they speculated on the nature of its internal workings. To Descartes it was contained in a cycle of Aristotelian infinity in which these dynamic mechanical processes were actually more important than the timescale in which they operated. These thinkers did not really attempt to determine the duration of earthly time, but left this question in a state of openness that reflected the ideas of Aristotle many centuries earlier.

The second strand emerged when, in contrast, many learned and religious men who were familiar with the landscape around them began to explain it and the Earth in the light of the dynamic and often catastrophic events described in the Scriptures. They accepted

Figure 3.1   Thomas Burnet (1635–1715) (from the fifth edition of his *Sacred Theory of the Earth* (1722), in Davies, *The Earth in Decay*, (1969), Plate 1).

that this had taken place over a very short timescale. God created the Earth and the organisms living on and in it, and so it seemed to these commentators that the wonders of his creation deserved to be examined closely and revealed to them. By the late 1600s several accounts of the world had appeared in print in England and in Europe which attempted to explain the nature of the Earth and its surface features with reference to the biblical readings. Those by Thomas Burnet (1635–1715) (Figure 3.1), William Whiston (1667–1752) and John Woodward (1665–1728) were the most widely known, with Burnet's ideas being particularly influential in some quarters and generating considerable comment and variants on the theory. On the other hand there were many in continental Europe who had serious doubts about these biblical-based 'sacred theories of the Earth' and argued that it was much older than these theories or the Bible allowed.

CARTESIAN AND KIRCHERIAN MODELS OF THE EARTH

Sweden in winter at the best of times can be extremely cold, but today, fortunately, few die of hypothermia in that country. In 1650, on the other hand, sitting in a Swedish castle at 5 o'clock in the morning discussing philosophy was a different matter, and in the case of René Descartes it was a serious matter: it led to his contracting pneumonia which quickly killed him.

René du Perrot Descartes was born on 31 March 1596 in the town of La Haye near Tours in France. Today his birthplace is styled La Haye-Descartes in his honour. He became the leading mathematician and philosopher of his generation: in fact his influence lasted far longer than his lifetime and extended way beyond the boundaries of his native France. He was fond of his bed, and did much of his thinking and writing while warmly wrapped up. At an early age he accepted the views of Copernicus concerning the Solar System and set out to write a book on the nature of the Universe. However, this was soon aban- doned when he heard that Galileo had been strongly criticised by the papal authorities for a similar book. Descartes entered the French army which provided a secure income and, because he was an officer, plenty of time to think. On his discharge he settled in the Netherlands where he married and had two daughters. His most famous book was *Discours de la méthode pour bien conduire sa raison*, published in 1637, in which he wrote the renowned line, 'I think, therefore I am'. Starting from a position of self-doubt, Descartes established a mechan- ical methodology which he applied to all of the Universe as well as to the workings of the human body, which he showed could easily be explained in purely mechanical terms. Many of his ideas are now referred to as being 'Cartesian' after the Latin version of his name, *Renatus Cartesius*. In mathematics he introduced the familiar square root symbol $\sqrt{}$ and the use of x and y in algebraic equations and in coordinates used to plot positions in three-dimensional space. In 1649 he was persuaded to travel to Sweden where he was welcomed into the court circles of Queen Christina. It was she who press-ganged poor René into early morning philosophical discussions that caused his health to fail and ultimately led to his death in Stockholm on 11 February 1650. Possibly Descartes' discussions had an effect on the eccentric Queen Christina, for she refused to marry, then abdicated her throne in 1654, converted to Catholicism, was eventually granted a pension by the Pope, and died in 1689. Descartes' head was detached from his body and the body was buried first in the Adolf Fredrik Kyrkogård in Stockholm, but later reinterred in Paris at the church of Saint-Germain-des-Prés. His skull, it has been said, turned up at an auction about a hundred years later, and was returned to France where it eventually came into the possession of the celebrated anatomist Baron Cuvier in Paris. Phrenologists who examined it noted that the anterior and superior regions of his skull were rather small. In this part of the brain is the cortical organ where, it was believed, rational thought occurred. The German Johann Gaspar Spurzheim (1776–1832) suggested that this small size clearly indicated that Descartes was not such a great thinker as had previously been believed. Today Descartes' skull is cared for in the Musée de l'Homme at the Palais de Chaillot in Paris alongside the other 35,000 human skulls in the collection.

Descartes' *Principia philosophiae*, published in 1644, does not contain much in the way of geology exposition but what he did write had a major influence on European writers on the subject for quite some time following. He described the Earth as having been formed from material derived from an extinguished star and having settled out in layers as it cooled. This theory of planetary formation, which broadly resembles what astronomers today think happened after the Big Bang, was first expounded in 1633, but only published after his death. Descartes thought that planets formed through the gravitational pull of a star which concentrated matter in space until it condensed from a gas, cooled, and formed planets.

The planet, or to be more precise the Earth, developed a distinctive internal layering as it cooled in four phases from the outside in. The whole planet was surrounded by an atmosphere (his layer B), while a dense crust of solid material (E) is underlain by a liquid layer (D), which itself overlies a hard layer of corpuscles, formed of small accreted spheres. The inner two layers consist of coagulated sunspot matter (M) surrounding a 'core' of hot corpuscles (I). As gases escaped from internal layers, cavity structures were left. Later collapse of layer E into these cavities created elevated mountains and depressions or basins which became the oceans, to be filled with waters escaping from layer D. In some earlier works Descartes had speculated on the origins of streams: they were, he thought, produced when subterranean waters that had seeped into the ground from the sea were evaporated at depth, and the water vapour rose through fissures in mountains towards the surface, where it condensed and emerged as springs.

While to some this may seem rather far-fetched, the vital aspect of Descartes' model for the formation of the Earth was that it was based on mechanical and physical processes which took into account the properties of the materials involved, and were governed by the laws of nature (or in modern terminology, physics). There was a logic to the Earth's formation. At no point did the hand of an external deity play any part in the process. While we do not know how long Descartes allowed for the formation of such a planet under his model, it was certainly episodic, and possibly quite short. He believed that geological processes could take place quickly. But time in any event was unimportant to him, just as it had been to Aristotle, and in this he differed markedly from those biblical chronologers to whom time and duration meant everything.

Like Descartes, Athanasius Kircher was educated by the Jesuits, but differed from him in that he chose to join the order. Born at Fulda in northern Germany on 2 May 1601, the feast day of Saint Athanasius, he had an inventive and fertile mind, and was considered to be one of the leading scholars of the seventeenth century. He was a professor at Würzburg and Avignon, and also studied and taught in Rome where he was revered and called the *Doctor centum artium* – 'the teacher of a hundred arts'. Among his exploits were his ideas on disease and epidemics, which he considered to be caused by microbes, an idea proved two hundred years later by Louis Pasteur. Although he devised his own rudimentary microscope, Kircher could not see the microbial disease carriers on the food that he had left to rot in his kitchen; Antoni van Leeuwenhoek's higher-quality instruments were not readily available at that time. He attempted to decipher Egyptian hieroglyphics but did not have the benefit of the Rosetta Stone, which still lay hidden in the waters near Alexandria. Kircher developed a magic lantern for projecting images on to walls (it is purely coincidental that some three hundred years later a Catholic priest in western Ireland used a similar device to project images of Our Lady on the gable end of his church. These 'apparitions' led to the site becoming a place of pilgrimage and an international airport can be found close by in the middle of a peat bog). He was also interested in water-powered church organs, and attempted to measure temperature by examining the buoyancy of small heated balls. Towards the end of his life he wished to become a missionary in China but was refused permission to embark on this calling, and he died on 28 November 1680 in Rome where his museum can still be seen in the Roman College. His heart was buried in a church he had built on the Sabine Hill, which today is a major place of pilgrimage.

Kircher's ideas on the Earth were first developed when he travelled through the volcanic regions around Naples and Sicily in 1637 and the following year, which coincided with a major earthquake that destroyed the town of Euphemia. Kircher was keen to see the effects of this event for himself and so travelled to Naples where he was lowered into the crater of Vesuvius from a rope. Using the pantometer, an instrument that he had invented himself for measuring angles, slopes and heights, he measured the dimensions of the crater. His ideas on the internal structure and dynamic nature of the Earth were developed over a long period, and only published nearly twenty-five years later in his *Mundus subterraneus* in 1665. His theories were, unsurprisingly for a Jesuit priest, in harmony with the biblical teachings, and he failed to speculate on the age of the Earth. However, his ideas were important as they were among the first to describe what happened within the Earth. Understanding these internal workings was an essential prerequisite to the understanding of how various natural phenomena arose. In this, Kircher more closely resembles many modern-day geologists who are interested in the dynamics of the Earth, rather than being more descriptive in their methodology. Kircher considered that the Earth was at the centre of the Universe. Internally it contained two systems of channels: one, his 'hydrophylacia', was water based, while the other, his 'phyrophylacia', was fire based. When the waters and fires interacted, earthquakes and volcanoes were produced, and springs emerged on the surface. He recognised that the molten magma he had seen at Mount Etna in Sicily was produced by heat at the Earth's centre, and that ores and sulphurous deposits were also formed through internal processes.

SACRED THEORIES OF THE EARTH

Descartes' and Kircher's ideas, which emphasised the mechanics of the Earth's system, were in stark contrast to the numerous biblically based theories that were to emerge in the late 1600s in England. Many of these, as Roy Porter pointed out in his historical volume *The Making of Geology: Earth Science in Britain 1660–1815* (Cambridge University Press, 1977), were very individualistic, and most of the authors did not make reference to the work of others. Again, these theorists were not particularly concerned with the age of the Earth: indeed, a number of authors criticised the chronologers for attempting to calculate a date for the Creation. As Thomas Burnet noted, it was difficult to be precise as there were three versions of the Bible and 'the most learned men are not yet able to determine with Certainty, which of the three Accounts is most authentick'. The theories were more concerned with explaining the nature of the Earth with reference to biblical events. Of these the Creation, the Flood and the impending Conflagration outlined in the Book of Revelation were the most important, and were those around which geological phenomena were explained.

Burnet's Telluris

Thomas Burnet, an Anglican priest, is best known for his work *The Sacred Theory of the Earth*, which he first published in Latin in two volumes in 1681 and 1689 as *Telluris Theoria Sacra*. Volume one contained two 'books' that dealt with the Deluge and Paradise, while the second also comprised two 'books' that were concerned with the conflagration and the emergence of a new Earth. He was encouraged to produce an English language translation, which duly appeared in two volumes in 1684 and 1689.

Burnet was born in the small village of Croft in Yorkshire and educated at the Free School, Northallerton, and Clare College, Cambridge. He later transferred to Christ's College and obtained a Fellowship in 1657. He was appointed Senior Proctor at Cambridge in 1661 and later served as tutor to the Duke of Bolton and the Duke of Ormond, James Butler, the premier Duke in Ireland. As is clear, he moved in high circles and appeared to be destined for a higher calling in the church: he was well connected in that his former college tutor was Archbishop Tillotson. Whatever ambitions he may have harboured, his chances of promotion were scuttled when he published a volume entitled *Archaeologia philosophicae: sive doctrina antiqua de rerum originibus* in 1692. He aroused very antagonistic views among fellow members of the Court and the Church because he treated the biblical account of Adam's sin and the fall of man as a fable. He was removed by King William from his position at court as Clerk of the Closet, and returned to Charterhouse College where he was Master. Burnet continued to publish but his later works did not attract as much attention as his earlier writings. He died in Cambridge on 27 September 1715.

While Burnet was influenced by the ideas of Descartes, he attempted to reconcile his theory with the biblical texts. The frontispiece to both the Latin and English editions, reproduced here as Figure 3.2, encapsulates his ideas regarding the Earth's evolutionary history. It shows a number of spheres that appear to be floating in space surrounded by angelic cherubs. Standing above the globes is a typically Renaissance representation of Christ, holding a flag in his right hand, and over him are the words in Greek 'I am the Alpha and the Omega'. The seven globes each represent a stage in the evolution

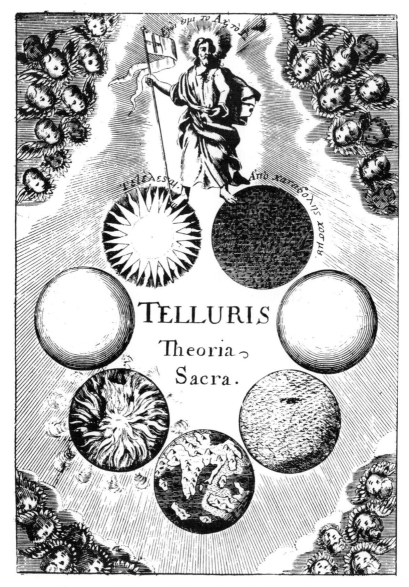

Figure 3.2   Title page of Thomas Burnet's *Telluris Theoria Sacra* (1681).

of the Earth, starting with that at Christ's left foot which shows the globe in a state of chaotic fluidity. As the Earth cools it reaches the second stage where the surface is smooth. In the third stage the Flood waters appear and a tiny wooden Ark can be seen floating across them. In stage four the present distribution of the continents and oceans can be seen. Stages five, six and seven represent the future. The Earth is consumed by fires of the Conflagration associated with the second coming of Christ in stage five, and in its next phase is restored in the Millennium to its perfect state as a smooth globe. In the final stage the Earth becomes a star.

The major aspect of Burnet's work was a computation of the volume of water present during Noah's Flood. Where had the waters come from to cover all the Earth to a depth of 15 cubits – a considerable volume? He estimated the volume of water required to achieve this level of flooding, and suggested that it was eight times the volume of water in the present oceanic basins. He illustrated how the volume of water could be estimated by taking soundings using a weighted line, but seriously underestimated both the areal coverage of the present oceans and their average depth. Nevertheless such mathematical frailties do not affect his resultant notions. But what, if there had been no topographic relief as is now present between the great ocean depths and the highest mountain ranges? If the Earth were smooth before the Flood, Burnet reasoned, then much less water would be needed to cover the surface. In fact, Burnet postulated that the same volume as is in the modern oceans would have been sufficient. Where had this water come from originally? Burnet lived with the inclement English weather, and realised that even continuous rainfall over forty days and forty nights would not have contributed a great deal of flood-water. Therefore the water had to have been derived from within the Earth. He invoked a scenario that had the Earth splitting open to shed its internal water, and believed that rocks of the surface crust then collapsed into the resultant voids. This produced the mountainous continents along the cracked margins and the ocean basins between, but there was still enough space for the Flood waters to return back into the Earth whereupon the dry land began to reappear. He realised that in some parts of Europe the seas were retreating or regressing and took this as evidence of the still dynamic process. Bogs and swamps with their waterlogged sediments were additional evidence that this event was rather recent. His ideas were quirky, but true to his religious beliefs. Apart from showing the Ark he also mapped the position of the Garden of Eden, which he represented in pictorial form with a clump of four trees situated near the equator in the southern hemisphere.

Burnet's *Telluris* received both praise and criticism: the Bishop of Hereford tartly remarked of Burnet that 'either his brain is crakt with over-love of his own invention, or his heart is rotten with some evil design.' It was widely read in Europe where a German translation was produced in Hamburg in 1698, and a third edition in Latin was printed in Amsterdam in 1699. Newton argued that the mountains were produced during the initial formation of the Earth, and also, as Stephen Jay Gould has noted, came up with the ingenious suggestion that the early Earth may have rotated on its axis much more slowly than it does at present. This would have produced a much longer day, the inference of which is that there was much more time than originally imagined in the six days of the Creation, and so plenty of time to form the Earth, its topographic features and its biologically diverse inhabitants. Others noted that Burnet's smooth second-stage globe was at variance with the Bible which read that the mountains and oceans were formed early in the six days and certainly before the Flood.

Burnet attempted to deflect the criticism of his ideas in several pamphlets published in 1690: in his *Review of the Theory of the Earth*, and in two small pieces, of 86 and 42 pages, which considered the matters raised by a Mr Erasmus Warren. In it he admitted that he did not know the age of the world, but suggested that it was no more than 6,000 years old.

John Woodward's dissolving Earth

Burnet's thesis was read by John Woodward, Professor of Physic at Gresham College, who felt it was lacking in original observations.

Being an experienced field geologist, Woodward considered that he could do a better job than Burnet, and so was spurred into action. His resultant account of the Earth and its evolution was entitled *An essay toward a natural history of the Earth; and terrestrial bodies, especially minerals: as also of the sea, rivers, and springs. With an account of the universal deluge; and of the effects that it had upon the Earth.* This 277-page volume appeared in 1695, and in subsequent editions in 1702 and 1723. An edition in French was published in Paris in 1735.

Although born in 1665 of humble Derbyshire stock, Woodward was able through his considerable intelligence to shake off this early burden, and carve out a career first in medicine and later in science. He was arrogant (perhaps his origins explain this disposition), and had an unfortunate ability to attract enemies as a magnet attracts iron filings. He did not like his opinions to be questioned by other men of learning such as his contemporary Fellows of the Royal Society, quite possibly because it may have suggested to him that they were of better lineage than himself. Indeed his outbursts led to him being expelled from the council of the society on at least two occasions.

By all accounts he was successful, well known for his publications and fiery character, and is today often recalled by British and especially Cambridge earth scientists. On his death in 1728 he asked that some land be purchased which would generate an income of £150. This money was to be given to the University of Cambridge to found a lectureship in geology and to pay for the curation of his extensive geological collections. Today a professorship carries his name, and he lies buried in Westminster Abbey.

Woodward was a avid collector of fossils (which he recognised were organic in origin), other geological materials and living biological specimens, many of which can still be seen in their original cabinets in the Sedgwick Museum in Cambridge. He published a pamphlet in 1696 that contained guidelines on how one should collect and preserve such material. This was the first curatorial manual and contained, for example, useful comments on dealing with preserving modern crabs or lobsters: 'chuse those [shells] that have the Creatures still living in them (which yet ought to be pluckt out, or they will putrifie and stink)', and preserving reptiles or small birds in 'small Jarrs, filled with Rum, Brandy, or Spirit of Wine, which will keep them extremely well.' Three hundred years later much the same methodology, but with industrial alcohol or formaldehyde, is used to preserve many biological specimens and Damien Hirst's sharks, pigs and sheep. (Brandy is usually not wasted today for such purposes, but drunk or used in brandy butter to complement and enhance one's Christmas pudding.)

Later, in 1728, the year of his death, his general thoughts on fossils were brought together in a volume entitled *Fossils of all kinds, digested into a method suitable to their mutual relation and affinity*, and in the year following his death a catalogue of his collection was published.

How did Woodward's ideas on the Earth differ from those prevalent at the time? A unique tenet of his thesis was the physical characteristics and dynamics of the sediments which he imagined had been produced when an 'old' Earth had become dissolved by the waters of the Flood.

The topography of his 'new' post-Flood Earth was formed as internal waters flowed from cracks and streams onto the surface. There sediments as well as organic remains settled out in distinct layers that circled the globe, and became lithified as they dried out under the effects of an internal heat source. Woodward imagined that the sediments settled out according to their specific gravity so that the densest settled first and the lightest last, and he noted that many of these layers contained fossils.

Today, anyone with some knowledge of the Earth's surface geology knows that there are many different rock types of varying densities at the surface. Granite is relatively light, while basalt is considerably denser; sand and sandstone, and lime and limestone also have a lowish density; and yet all these rocks in parts of Britain and Ireland can be found within a short distance of each other cropping out on the surface. Woodward was clearly incorrect in his theory of the

Earth's structure, but it is easy to understand how he could have made this mistake. If you take a glass beaker and pour into it a couple of handfuls of sand collected at a beach, then fill the container with water and agitate it violently, when set down the heavier sediments will settle out. However, the pattern may become somewhat cloudy if you have sediments of equal density but of different grain size within your container. Then the coarser sediment will settle first. One can imagine Woodward carrying out such experiments, but although he was an accomplished field geologist he failed to understand the message the surface rocks were giving him. Additionally, fossils are not arranged in a sequence according to their density but in a biological lineage that is better understood now than in the late 1600s, when the biological affinities of fossils were not accepted by all.

Nevertheless, as Martin Rudwick has pointed out, Woodward's ideas were important as he attempted to give reasons for the presence of characteristic fossils in an ordered succession of strata. Further understanding in the early 1800s of this palaeontological characterisation of the rock succession played an important role in the development of stratigraphical geology and biochronology.

The confrontational Whiston

One of Woodward's minor critics was William Whiston, who produced his own theory of the Earth in 1696, translated into German in 1713. The 1696 volume carried the splendid, long and explicit title: *A new theory of the Earth, from its Original to the Consummation of All Things, wherein the Creation of the World in six days, the universal deluge, and the general conflagration, as laid down in the Holy Scriptures, are shewn to be perfectly agreeable to reason and philosophy.* In the title alone Whiston nailed his colours to the Mosaic mast. He followed this up two years later with a volume containing a vindication of his earlier book and its contents.

William Whiston was born in 1667 in Leicestershire in the village of Norton where his father served as the Rector. A sickly child, he was educated at home. He later entered Clare College, Cambridge, where he demonstrated great mathematical ability and was elected a Fellow in 1693. Later he was given the living of Lowestoft in 1698, a position he combined with being chaplain to the Bishop of Norwich. He resigned shortly afterwards and returned to Cambridge where he succeeded Sir Isaac Newton as Lucasian Professor of Mathematics. Unfortunately for him he published a work on the foundations of the church and his unorthodox views led to his being stripped of his professorship and expelled from the university. At a church service in 1747 he stood up, walked out and severed his links with the Church of England, and then allied himself to the Baptists. He remained a confrontational figure for the remainder of his life, which ended at his son-in-law's house in London on 22 August 1752.

Whiston's main criticism of Woodward's ideas lay in Whiston's refutation of the notion of miracles. But Whiston added more, and claimed a natural cause for the Flood: this he said had been caused when a comet passed close to the Earth and condensation of the vapours in its tail had been sufficient to cover the Earth with water. He also noted that the sediments that were supposed to have been precipitated out of the floodwaters were not always laid down in the order of decreasing density. Whiston's book was described in 1911 as being 'destitute of sound scientific knowledge' but it at least brought 'some new things to our thoughts'.

For many, such as the English letter writer John Locke, the biblical chronologies were just too short, and the shoe-horning of sacred theories of the Earth into this limited span of time, coupled with the use of physical parameters determined by the biblical story, was simply not credible. These sacred theories represent an excursion into a cul-de-sac of geological fantasy, sprinkled with crumbs of geological observation, which today do not stand up to serious scientific scrutiny. However, they were important as they were early attempts to explain the complex geological and evolving nature of the Earth, a dynamic system which was, these early authors believed, constrained by a short timeframe.

4 Falling stones, salty oceans, and evaporating waters: early empirical measurements of the age of the Earth

In the latter part of the seventeenth century in England and on the Continent there emerged scholars prepared to argue against the various sacred theories of the Earth and to discuss the possibility that the Earth was older, possibly immeasurably older than these theories would allow. The link to the biblical texts was weakening and two men in England, in particular, demonstrated or suggested that scientific observations and scientific experimentation could produce empirical data from which to deduce a number in years for the age of the Earth. Today one of these men is little known even within scientific circles, whereas the second remains very well known on account of his prediction of the timing of a returning celestial object.

EDWARD LHWYD

Oxford has long been a centre of learning, with its many colleges, spires, dons and students. Today, students bicycle at speed from lodgings to lectures, dodging tourists and thinking about their next meal or assignation. On Cornmarket Street stands the small church of St Michael at the Northgate. Here, as we shall see, one might expect to find a memorial to the first geochronologist that has attracted my attention. This church, the Saxon tower of which dates from the eleventh century, is the oldest building in Oxford. How many cycle past and never enter its doors? Probably hundreds every day. If any one of these stopped, parked his bicycle by the entrance and walked inside he would find no fitting memorial. Instead, he would need to remount his bicycle and make his way to the splendid masterpiece of Victorian Gothic architecture, the Oxford University Museum of Natural History

Figure 4.1 Edward Lhwyd (1660–1709) (from R. M. Owens, *Trilobites in Wales* (Cardiff: National Museum of Wales, 1984), p. 4). Courtesy of the National Museums and Galleries of Wales.

on Parks Road. There, through the rather unassuming front door, can be found the remains of the geological collections of Edward Lhwyd (Figure 4.1), fossils that he illustrated in a now classic pocket-sized book published in 1699 (see Figure 4.2).

Edward Lhwyd, whose name has been spelt in a numbing variety of ways: Floyd, Lloyd, Llhwyd, Lhuyd, Llwyd, Luid, Fluid, or in Latin, Luidius (he signed himself as 'Lhwyd' in correspondence, so that is the form used here) was a man of numerous interests – natural history, Celtic philology, antiquities, to name but three – who travelled widely in Britain and Ireland at a time when such trips must have been both time-consuming and arduous. Born at Glan Ffraid (or Llanvorda) near Oswestry in Shropshire in 1660, he spent much of his childhood living on his father's estate nearby. His father Edward Lloyd was unfortunate in being a Royalist at the time of the Cromwellian disputes, and consequently lost much of his estate, becoming destitute. His mother, Bridget Pryse of Gogerddan in Cardiganshire, with whose family he frequently spent his holidays, had the stigma of bearing Edward as an illegitimate child. Edward senior died young, and it is all the more remarkable that his son managed to get a good education. Edward, or Ned as he liked to be called by his friends, attended Oswestry Grammar School and then entered Jesus College, Oxford, in 1682 to study law. Jesus College, which received its Royal Charter in 1571, had been established through a bequest by Dr Hugh Price of St David's Cathedral, and had strong Welsh connections, which it still retains. One relatively recent alumnus was T. E. Lawrence, better known as Lawrence of Arabia, who graduated in 1910. Lhwyd remained at Jesus for five years but did not graduate with a degree. He became side-tracked by a new interest, one that was ultimately to supply him with a modest living, and that allowed him to exercise his consider-able imagination and skills as a collector of facts and data – skills that probably would have served him better in the Middle Temple. Little is known of his character and only one portrait exists: a small black and white ink drawing that exists in two known versions, one deco-rating a document now in Merton College, Oxford, the other in the Ashmolean Museum. One version has Lhwyd's face framed by the first initial of his Christian name; the second, which is reproduced here (Figure 4.1), is clearer as this initial has been removed. In this half-length portrait he is wearing a gown, a white scarf is wrapped tightly around his neck, and his head is covered by a wig of shoulder-length white curls. He has striking eyes, and a slight smile which hints at his good sense of humour. Probably drawn in his late thirties the portrait suggests that he enjoyed the meals served up at the College refectories, as evidenced by the soft rounded features around his chin and neck.

ROBERT PLOT'S ASSISTANT

Soon after he arrived in Oxford, Lhwyd began to assist Robert Plot, Professor of Chemistry and Keeper of the Ashmolean Museum from 1683 to 1691. He was subsequently appointed his assistant and as 'Register of the Chymicall courses at ye Laboratory', positions that must have carried a small but necessary stipend. This institution was housed in the first purpose-built museum in Britain, which was erected between 1679 and 1683 beside the Sheldonian Theatre,

Christopher Wren's first major commission. Today the building still serves as a museum, where it houses the collections relating to the history of science: there amongst the displays you can find Lewis Carroll's camera and Albert Einstein's blackboard. The impressive entrance to the Old Ashmolean Museum is flanked by paired Corinthian columns and reached by a number of steps cut in Portland Limestone from southern England. The surrounding wall is protected by busts of four Roman emperors, with a further thirteen surrounding the Sheldonian. These are occasionally misidentified by mathematically challenged tourists as representing the twelve disciples. One can imagine Lhwyd's excitement when he entered the doors on his first day to help out Plot, and later as a paid employee of the University. Once inside the building, he would have made his way to the upper floor exhibition galleries where he might have found Plot poring and puzzling over some petrifactions, or the collections of the Tradescants father and son, or of Elias Ashmole. Later, on his induction tour, he would have stepped inside the ground floor lecture theatre and perhaps measured his height against the lectern, before descending beneath ground level to the basement laboratory, which was equipped with fancy chemical apparatus. These rooms would in time become a second home to him, even though he kept a home a number of miles outside the city at Eynsham.

Robert Plot (1640–1696) was a Kent-born naturalist and chemist who was appointed first Keeper of the Ashmolean in 1683. Today much of his fame, or infamy, lies in his authorship of two volumes of observations on natural history and antiquities. His first book *The Natural History of Oxfordshire* was published in 1677 while his second *The Natural History of Staffordshire* appeared nine years later in 1686. He was, by all accounts (and there are not many), a rather odd man, but nevertheless a good and careful curator. However, in 1690, at the height of his fame he resigned his chair and keepership citing the low salary as good reason for leaving: it was better, he said, to do something rather than sitting around doing nothing for nothing. He retired with his new wife to Sutton Barne, where he had property, but enjoyed only six years of marriage before he died following complications with his urinary system.

The earlier reference to Plot's infamy derives from his interpretation of various English fossils which he described and illustrated for the first time in his two county books. Today young students of palaeontology find these accounts fanciful, amusing, even startling. Plot described 'screw-stones', 'bulls' hearts', 'horses' heads' and 'star-stones' as well as a variety of other petrifactions, now known to be the fossilised remains of once living marine organisms. He attributed their formation to some 'plastic force' and did not regard them as being of organic origin. His 'screw-stones' are either Lower Carboniferous crinoid stems or the internal moulds of turritellid gastropods; his 'bulls' hearts', or *Bucardites* as he called them, are the internal mould of the bivalve *Protocardia*; the 'horses' heads' or *Hippocephaloides* are internal moulds of another bivalve, *Myophorella hudlestoni*, from the Jurassic rocks at Headington near Oxford; and his 'star-stones' are colonial corals. However, before condemning Plot for his ludicrous conclusions, one should look at the material: when viewed from a particular angle, some of the internal moulds do look like horses' heads or hearts. With nothing with which to compare the material, it is not surprising that he reached the conclusions and attributions he did. Plot was important in the history of palaeontology as he brought this material to the wider public and thus instigated a debate on the true nature of the material, this at a time when the organic origin of such curiosities was beginning to be appreciated across Europe. He also illustrated in his *Oxfordshire* treatise what he thought was the petrified thigh bone of a giant man. Nearly ninety years later in 1763, Richard Brookes redescribed Plot's specimen, which he named *Scrotum humanum* – no guessing what portion of the human he thought it represented. Today, we know that this bone is the lower portion of a thigh bone from the dinosaur *Megalosaurus*. This animal, first described by the later Oxford academic William Buckland, was the first dinosaur to be given a name. Recently, it was suggested by the palaeontologists and historians of science Bill Sarjeant and Beverly Halstead that in fact

Buckland's name was invalid, and that Brookes' epithet *Scrotum* had priority as the generic name of this dinosaur! Sadly the International Commission for Zoological Nomenclature ruled in favour of *Megalosaurus* and *Scrotum humanum* is considered to be a *nomen dubium* which has now been consigned to the waste paper bin as a nomenclatorial oddity.

A FOSSIL CATALOGUE AND THE FALLING STONES OF LLANBERIS

Let us return to Edward Lhwyd. In the first three years of his keepership he spent a great deal of time curating and cataloguing the geological specimens in the Ashmolean, and this work was eventually published as *Lithophylacii Britannici Ichnographia* in 1699. This small book, octavo in size, designed to be portable and, as such, one of the first field guides ever produced, contained 139 pages and 23 plates. He had difficulty getting it published, as the usual publishers to the University claimed that it was not viable, so he turned to patrons and subscribers including Isaac Newton and Hans Sloane who came up with the funds that allowed him to produce an edition of only 120 copies. It is essentially a catalogue of 1,766 specimens, but also contains a number of useful appendices in the form of letters, in which Lhwyd discusses the nature of fossils. He actually believed that they were formed from seeds that had been blown or washed into cracks in the rocks. The book was popular and a pirated edition appeared in Germany later that year. A copy of this work inscribed by the author can be found in Trinity College Dublin, and while some of the Latin text is difficult to decipher, many of the fossils engraved are clearly recognisable. He illustrated Upper Carboniferous plants, dinosaur teeth and trilobites among a suite of other fossils (Figure 4.2). Some of the trilobites had been described by him earlier in a paper to the Royal Society: these 'flat-fish' are now recognised as the common species *Ogygiocarella debuchii* found in the Ordovician successions around Builth Wells in central Wales. Unfortunately when the 1699 edition was being typeset Lhwyd was away from Oxford and it is

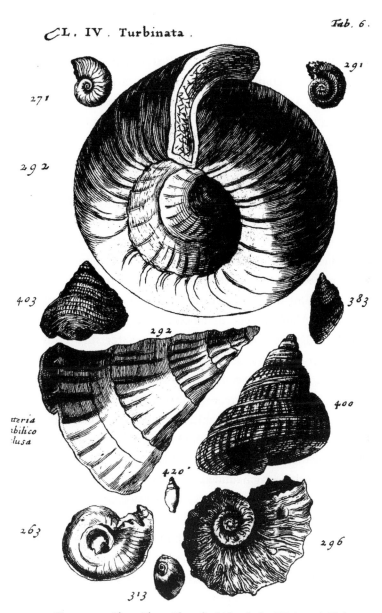

Figure 4.2 Plate 6 from Lhwyd's *Lithophylacii Britannici Ichnographia* (1699), showing various fossil cephalopods and gastropods (Geological Museum, Trinity College Dublin).

littered with typographical errors. These were corrected in a posthumously published edition in 1760.

Lhwyd's specimens did not fare well after his death. His favourite pupil David Parry, companion on many of his travels, and successor as Keeper at the Ashmolean, was neglectful of his duties and took to drinking, and the materials got lost or documentation became detached, rendering identification difficult. By 1945 it was stated that only two of the original suite of fossils could be recognised. Today, largely thanks to the diligent work of the late John Edmunds of the Oxford University Museum of Natural History, many more specimens can now be identified with certainty in their collections as having been catalogued in the *Lithophylacii*.

Once he had largely completed his catalogue, Lhwyd realised that he had more freedom, and the possibility of increasing his annual earnings; at no time did they go over £50 a year, and he complained that at times he had to seek assistance from his 'unkle'. He was approached by Dr Edmund Gibson who was revising Camden's *Britannia*, a sort of early gazeteer and topographical dictionary, and Lhwyd offered to cover three counties in Wales. With this commission he began in 1693 a long series of travels in Britain, Ireland and Brittany. He collected a great deal of information for Camden's and then decided to produce a multi-volume work on Wales, which would be written along the lines of Plot's Oxfordshire natural history. For this he gained sponsorship which funded the work – but it was tough. As he became more interested in antiquities, he developed a fascination for native languages and apparently coined the term 'Celtic'. He travelled in Scotland in 1699, Ireland and Cornwall in 1700, and Brittany in 1701 collecting manuscripts (many of which eventually found their way into Trinity College Dublin, where they are housed alongside James Ussher's library). In Ireland he collected and noted the presence of many rare plants and examined ogham stones in County Kerry – odd upright stones that carried an alphabet consisting of notches and lines carved into their corners – and generally gathered information. Some of this found its way into his *Archaeologia Britannica* which was

published in Oxford in 1707, but unfortunately the remaining proposed volumes never appeared. Travels were not without their difficulties. Usually he had three assistants with him, but one absconded
in Wales. In Cornwall the remaining trio were arrested in the town of
Helston where they were accused of stealing. Having managed to
extricate himself and his companions, Lhwyd probably felt that nothing worse could befall them. Nothing worse? In Brittany the following
year Lhwyd was arrested and charged with being a spy – a very serious
charge – but fortunately after a spell of ten days spent in prison at Brest
he was released on condition that he left France immediately. One
would imagine that he was happy to comply. He then returned to
Oxford and began to write up his travels.

In the context of this story, he was perhaps an unwitting con-
tributor to the later debate on the age of the Earth. As we have heard,
he travelled widely, particularly in Wales. At some point before 1691
he ventured up Snowdon, the highest peak in the northern mountainous region. He collected plants including the Snowdon lily, *Lloydia
serotina*, which today is very rare indeed in Britain with only five
patches clinging precariously to the slopes. Lhwyd was the first naturalist to note that Britain's higher peaks supported a distinctive
alpine flora. Today *Lloydia* thrives in the Alps, the Himalayas, in
Alaska and in northern Siberia.

The pass at Llanberis is regularly used by hikers in the region,
and allows relatively easy passage to higher slopes. Underlain by
Cambrian sediments and coarse intrusive granites and other igneous
rocks, the valley is steep-sided, flat-bottomed and most impressive
(Figure 4.3). Now certainly Lhwyd would not have appreciated its
origins as a glaciated valley – the recognition of past glaciers in Great
Britain and Ireland did not come until 150 years later with the visit by
Louis Agassiz in 1840 – but he was intrigued by the large boulders that
lay on the valley floor, and on that of the adjacent valley of Nant
Ffrancon. On 30 February 1691 he wrote to his friend John Ray
(1627–1705), who had earlier written about erosion on the surface of
the Earth and who found it difficult to imagine that geological time

Figure 4.3   Llanberis Pass, north Wales. This valley, produced by glaciation, cuts through various igneous bodies. Scattered erratics lie on the slopes and valley floor particularly in the middle distance. Photograph by J. Rhodes, 1934. By permission of the Director of the British Geological Survey.

was short, particularly when he had observed marine shells exposed high above the present sea surface. Ray was a frequent correspondent, and published a list of plants Lhwyd had collected around Snowdon in his *Synopsis Methodica* in 1690. Lhwyd described how he had seen numerous boulders in the two valleys: 'there are but two or three that have fallen in the memory of any man now living, in the ordinary course of nature we shall be compelled to allow the rest many thousands of years more than the age of the world.'

Here Lhwyd was demonstrating direct evidence that the Earth was much older than Ussher, Burnet, Whiston or Woodward had thought, or even dared think. Ray reported this in the second edition of his *Miscellaneous discourses concerning the dissolution and changes of the World* in 1692, but did not exactly agree with Lhwyd, whose logic he found difficult to follow. Perhaps the boulders were thrown down the mountain slopes by the actions of the Flood? Lhwyd, who violently opposed the sacred theorists, could not subscribe to this viewpoint, like Ray and an ever-increasing number of other men of science who were distancing themselves from the earlier theories of the Earth. It is easy now to recognise Lhwyd's conclusions as being severely flawed, knowing as we do that the boulder fields at Llanberis were formed as huge boulders were deposited by melting ice during the Pleistocene ice age. However, if one applies figures to Lhwyd's observations what age do the boulders indicate? Most adults at the time lived until their sixties, and if two or three boulders fell in that time, then one fell every 20 to 30 years. A count of the boulders suggest that there are at least 10,000 at Llanberis, which gives between 200,000 and 300,000 years as the age of the Earth using Lhwydian logic. This may not sound long, but in the context of biblical chronologies, it is fifty times as great.

Lhwyd was an enthusiast and a workaholic, who in his short life of 49 years made major contributions in linguistics, natural history and antiquities. He was elected a Fellow of the Royal Society, but not without some difficulty. His nomination had been opposed by John Woodward with whom he had clashed over their differing views of the nature and formation of fossils. It is ironic, given his early pecuniary circumstances, that Lhwyd was to die at a time when his financial situation looked to have eased through his appointment as a minor officer in the University. During the month of June 1709 when the battle of Poltava, which pitted the army of Charles XII of Sweden against the Russian forces of Peter the Great, was nearing its end, Lhwyd probably spent most nights sleeping, as was his habit, in his office or quarters in the Ashmolean. His rooms were damp and probably stuffy (three hundred years later similar conditions are still, unfortunately, frequently found in museums). Staying overnight in such conditions, together with his life-long affliction with asthma, combined with pleurisy, did little for his health, and he died in the Ashmolean on 30 June 1709.

If you happen to visit Oxford and have the time, visit the ancient church of St Michael's and walk up the south aisle. This was associated with Jesus College and, appropriately given Lhwyd's Celtic interests, is known as the 'Welsh aisle'. There, somewhere beneath your feet, lie the remains of Edward Lhwyd, the chronologist who dated the Earth by counting the number of boulders at Llanberis.

FRESH AND SALTWATER

Our second inquiring geochronological mind belonged to Edmond Halley (1656–1742) (Figure 4.4), the Astronomer Royal, whose name is associated with the comet whose return every 76 years or so he predicted. A representation of the 1066 sighting is embroidered onto the Bayeux Tapestry, and it last passed our skies in 1986 to great popular excitement and frenzied scientific analysis. He is certainly well known today, particularly when compared with Lhwyd. Recently I typed both names into an Internet search engine, and while one might argue that the returns do not necessarily reflect a true measure, they do count for something. 'Edward Lhwyd' returned 91 hits and the variant 'Edward Lhuyd' returned 210 hits for a total of 301; other variants produced negligible returns. In contrast 'Edmond Halley' was contained in 15,250 sites, while 'Edmund Halley' returned 20,215 hits for a total of 35,465 sites altogether! Lhwyd's returns represent only 0.8% of Halley's. However, unlike Halley, Lhwyd is commemorated in some generic names applied to living organisms. The starfish *Luidia* was named for him by the naturalist Edward Forbes in 1839, and *Lloydia*, the Snowdon lily, recalls his memory and his travels in that district.

In 1715, just six years after Lhwyd's death, a short paper appeared in the pages of the *Philosophical Transactions of the Royal Society*, penned by Edmond Halley, Secretary of this now old and venerated scientific body. While not perhaps one of his most famous papers, it was certainly original. However, it suffered the fate of many papers in that the ideas it was promoting and attempting to propagate were soon forgotten and sank into obscurity. The paper and its ideas were only discovered and brought to light approximately 200 years

Figure 4.4 Edmond Halley (1656–1742) (from John Francis Waller (ed.) *The Imperial Dictionary of Universal Biography*, vol. 11 (Glasgow: William Mackenzie, *c.* 1857)). Courtesy of John Wyse Jackson.

after its publication. In terms of ideas on geochronology, it is an important work.

Edmond Halley, along with Sir Isaac Newton, ranks as one of the best-known English scientists of the seventeenth and eighteenth centuries. His father Edmond was a soapboiler who had extensive property interests in London, and who died a wealthy man reputed to be worth £4000 (£1.6 million in today's currency). Living in London, Edmond junior (born in 1656) survived the Great Plague in 1665 and the Great Fire in 1666 but his school, St Paul's, did not. In 1673 he went up to Oxford, where he studied at Queen's College. After graduating he spent a year on St Helena (later to find fame as Napoleon Bonaparte's final exile home), where he set up an observatory. In 1680 he observed a comet while travelling to Paris: this comet now bears his name. In 1686 he was appointed Clerk to the Royal Society and his *Principia*, which contained his theory of comet movements, was published the following year. By 1695 he was making further calculations about the trajectories and timings of various comets and mapping the positions of stars, and then in the next four years undertook a number of long scientific voyages in the Atlantic. He was elected as Savilian Professor at Oxford in 1704, having earlier failed to be elected Savilian Professor of Astronomy. The election coincided with an increase in his influence in scientific circles generally. Oxford awarded him a Doctor in Laws in 1710 and three years later he was elected Secretary of the Royal Society. In 1720 he reached the scientific zenith for astronomers in being appointed Astronomer Royal.

His wife Mary (née Tooke), whom he married in April 1682, died in 1736 after a marriage of 54 years – a union of remarkable longevity for that time – and she was buried in the church of St Margaret at Lee, near the Royal Observatory at Greenwich. The couple had three children: Edmond who predeceased his father by two years; Margaret who died in the year following her father; and Katherine, who survived both her father and her two husbands. Edmond Halley died on 16 January 1742 and given his stature, one might be forgiven for expecting that on a visit to Westminster Abbey one could find his tomb alongside that of other great astronomers, scientists and thinkers of the period, including Newton. Instead Halley was buried at Lee beside his wife, and in due course his two daughters were buried alongside in the same tomb. In the nineteenth century the church was rebuilt and the plaque that had covered the tomb was removed to Greenwich where it remains. Only in 1986, some 244 years after his death, was a small memorial placed to Halley in Westminster Abbey. Perhaps he needs no physical memorial on Earth; after all he is commemorated in name by the comet whose return he predicted.

Halley, like many men of science, turned his labours to a variety of subjects. In his case, while astronomy was his major field of enquiry, he also examined problems of navigation, the Earth's magnetic field, barometric pressure and the structure of the atmosphere (which he determined was layered), the distribution of trade winds and monsoons, and man's lifespan. He even invented a diving bell in 1691 in which he was lowered to a depth of 18 metres (10 fathoms). When sitting on the bed of the River Thames he was able to forget about the bubbling and heaving metropolis above, and

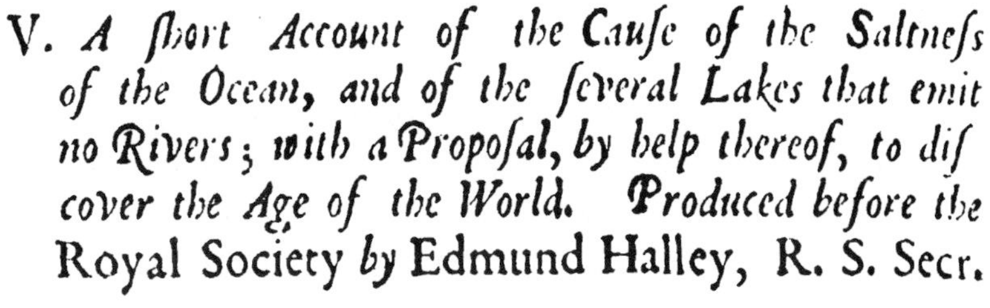

V. *A ſhort Account of the Cauſe of the Saltneſs of the Ocean, and of the ſeveral Lakes that emit no Rivers; with a Propoſal, by help thereof, to diſcover the Age of the World.* Produced before the Royal Society by Edmund Halley, R. S. Secr.

Figure 4.5   Title of Edmond Halley's paper on the salinity of closed lacustrine systems (from *Philosophical Transactions of the Royal Society* **29**, number 344 (1715), 296–300).

relaxed by drinking punch and smoking a pipe. He drew up plans for a diving suit, but it was not until two decades later that Andrew Becker's celebrated version was successfully tested, again in the waters of the River Thames. And, of course, he considered aspects of the physical nature of the Earth, its interior and its age. His St Helena years form part of our story of geochronology, in that while there, Halley became interested in the hydrological cycle where he observed water condensing from clouds as rain, which then made its way to the sea through rivers and streams. Eventually, through evaporation, water returns to the atmosphere where it forms clouds and the cycle can continue. This cycle he described in a paper of 1691. While all this sounds obvious to us, and now graces the pages of even the most rudimentary of geography textbooks, this cycle and the relationship of clouds to streams and springs was not well understood in the late 1600s.

In 1715 Halley suggested that the age of the Earth could be derived by examining the saltiness of lakes, and published his musings in a short paper entitled *A Short Account of the Cause of the Saltness of the Ocean, and of the several Lakes that emit no Rivers; with a Proposal, by help thereof, to discover the Age of the World* (Figure 4.5). In essence the thrust of the paper was as follows: if one was to measure the concentration of salt contained in lakes that lacked a river exit, over time one would find that the concentration would increase. This would yield an actual figure in years for the age of the lakes, assuming that they contained no salt when they first formed. This methodology could be expanded to measure the saltiness of the oceans, and this, he said, had implications for the age of the Earth. It was a clever idea, no doubt formulated on account of his interest in the hydrological cycle. However, it does have a number of flaws. It would be difficult to find closed lakes, although some do exist: Halley suggested that the Caspian Sea and Lake Titicaca could be examined. The largest such closed system is the Aral Sea in southern Russia, but today this is shrinking as more water is drawn from it than enters it. Halley's method did not take into consideration salts washed into the under-lying soils and sediments on which the lakes sat and therefore removed from the 'closed' system. The scheme would also take a long time: it was unlikely that instruments were precise enough at the time to be able to distinguish salt concentrations in two water samples taken from the same place ten years apart. Halley knew this and regretted that the Egyptians or Greeks had not measured the oceanic salt concentrations two thousand years earlier. Hedging his bets and being naturally somewhat cautious, given the power of the Church, Halley noted that his scheme would yield a maximum age for the Earth, but said that 'the World may be found much older than some have imagined'.

In 1724, in a further paper delivered before the Royal Society, Halley tackled the question of the Flood, or Deluge as it was often called at the time, and its effects on the Earth. Given his interests in comets, it was natural for him to invoke one to form mountains. When the Earth was struck by a passing comet, Halley argued, the collision caused the sea to move landwards, rather like modern-day tsunamis or giant tidal waves, carrying marine sediments which were deposited in great piles on the terrestrial surface. As the waters flowed back into the oceans the sedimentary piles were left behind as mountains.

Halley's ideas on salt-clocks and time were soon forgotten and only resurrected in 1910 by the American geochronologer George F. Becker, at a time when it was thought that salt and the oceans held the key to unlocking the secret of the Earth's true age. This episode is discussed further in Chapter 12.

Many late seventeenth and early eighteenth-century commentators remarked on the difficulties of estimating geological time. We have seen the empirical evidence presented by two Englishmen, but they were not alone in their efforts. On the Continent, and particularly in France, others were similarly doubting that the Earth was very young, but few attempted to quantify just how old. One of these men, a French diplomat and naturalist, Benoît de Maillet (1656–1738), who was interested in sedimentology and who had with his grandfather studied sediments and the disposition of fossils in them, wrote a book over a long period between 1692 and 1718 which he called *Telliamed; ou, Entretiens d'un philosophe indien avec un missionnaire français sur la diminution de la mer, la formation de la terre, l'origine de l'homme, &c.* He attempted to get it published but failed and was forced to circulate manuscript copies of the work clandestinely. Seven of these are still known to exist, but the whereabouts of the original is unknown. It created quite a storm among the learned men of Europe and in the hierarchy of the church, and was only finally published some thirty years later, after the author's death. The manuscript had been entrusted into the care of the Abbé le Mascrier who was so worried about the unorthodox religious content of the book that he modified portions of it, but also took the step of publishing it under the named editorship of a lawyer, Jean Antoine Guer, who in fact had absolutely nothing to do with it. Monsieur Guer's reaction to his association with the volume is unfortunately not recorded. By the third edition the abbot was confident enough to admit that he was its editor. Three editions appeared in French in 1748, 1749 and 1755, and two in English in 1750 and 1797. This last was printed in Baltimore and aptly contains in the title 'A very curious work'.

Maillet was of noble stock and was brought up in the region of Lorraine. At the age of thirty-five he became a career diplomat, first serving as Consul General in Egypt, and then from 1708 as Consul in Livorno on the northwest coast of modern-day Italy. After spending seven years in this area of beautiful and interesting geology, he became an inspector of French interests in the east and on the Barbary Coast of North Africa. All diplomats hope to retire on a good government pension and his became available in 1720. He spent two years in Paris before finally settling for the remainder of his life beside the sea at Marseilles.

Maillet was influenced by Descartes' ideas and did not accept the biblical accounts when it came to chronological matters. His thesis paralleled the Cartesian scheme in that the Earth was a former star or sun, but he added his own ideas: the Earth was once completely covered with water, and he argued that sea levels had dropped through time. To test this assertion he devised his own personal hydrographic station through which he measured changes in sea level over a considerable time-span. Like Leonardo da Vinci, Maillet had observed seashells high up on Italian mountainsides and recognised that they must have been underwater at some point in the past. As the sea level dropped, eventually the Earth would dry out and dessicate, and the internal volcanic fires would cause it to re-ignite and become a star again. Mountains, he suggested were formed on the sea bed of sediments piled high by strong underwater currents, and then exposed as the sea levels dropped. He also suggested an evolutionary sequence for plants and animals: higher plants such as trees were derived from seaweed, while all animals had a marine source too. Birds evolved from fish, terrestrial animals from marine animals. In this, he was broadly correct.

These thoughts were documented in the book in a very strange style: the narrative consisted of a conversation between two very different people, an Indian philosopher and a French missionary, and it could be argued that he did this in an attempt to have the work considered a piece of fiction. It was not. The philosopher, named Telliamed, which any devotee of crosswords will immediately spot is de Maillet spelt backwards, argued that as ocean water evaporated, the water vapour was lost into space, and consequently sea levels fell at a rate of three inches per century. This figure was based on his own hydrographic observations. On the basis of knowing the height above sea level of various seashells, he was able to get an estimate of the age of the Earth, which he considered to be over 2 billion years old. He also suggested that man had been in existence for four hundred thousand years. No wonder nobody would publish this work for many years! These conclusions were fantastic, and well beyond the comprehension of eighteenth-century readers and scientists.

It is now well documented that changes in sea levels have occurred in the geological past. These eustatic changes, as modern-day geologists term them, do not just occur in one direction. Sea levels have fluctuated markedly in the past: recent changes associated with the last Pleistocene ice age resulted in Ireland and Great Britain being joined to continental Europe for a period as sea levels fell, before rising sea levels separated them again, both from it and from each other. Maillet did not recognise that levels could fluctuate, nor did he realise that continental masses could, by various methods, be elevated out of the sea so that marine sediments could easily be found as a result at high altitudes. This is hardly surprising as these ideas were only formulated as late as the 1920s.

Certainly these early empirical age estimates of Lhwyd, Halley and Maillet for the Earth were wrong; but they were the first serious scientific attempts at precise geochronology and should be respected as such. Later, in the nineteenth century, other methods using sedimentation rates and revisitation of the salt-method followed, before the breakthrough in the twentieth century of radiometric dating.

5 Thinking in layers: early ideas in stratigraphy

The late Roman Catholic Bishop of Kilfenora, a tiny diocese in County Clare in western Ireland, was responsible on 23 October 1988 for beatifying another bishop and elevating him to the position of Blessed, the first step along the time-consuming path to sainthood. It would be unusual, one might think, for a minor bishop to have such power, but not when you discover that amongst the Bishop of Kilfenora's other sees is that of Rome. This bishop was none other than John Paul II, the late Pope, who had been responsible during his pontificate for creating more saints and beatifying more persons than did his five predecessors combined. In total Pope John Paul II canonised 469 people and beatified over 1,300 more. The subject of the autumn 1988 ceremony held in Rome was Nikolas (or Niels or Nicolaus) Stensen, the Titular Bishop of Titiopolitan, whom geologists and historians of science are more likely to remember by the Latin version of his name, Steno. Geologists revere him as the 'father of geology', and his ideas gave rise to what has been labelled by some 'the Stenonian Revolution' in geology.

STENO AND THE TUSCAN LANDSCAPE: TURNING THE KEY
TO THE CONCEPT OF STRATIFICATION
Nikolas Stensen (Figure 5.1) was born in Copenhagen, Denmark, on 20 January 1638 and died 48 years later in Germany on 25 November 1686. In that comparatively short time he achieved much in several fields: in medicine, natural history and in the church.

He trained at medical college and became a noted anatomist, publishing two celebrated treatises, that on muscles, *De musculis et glandulis*, in 1664 and *Discours sur l'anatomie du cerveau* in 1669. He was consecrated Bishop of Heliopolis in 1677. On his death, he left little in the way of material goods, and was buried over six months

Figure 5.1 Nicolaus Steno (1638–1686) (from J. G. Winter, *Prodromus of Nicholaus Steno* (1916), Plate 5).

later in the Basilica of San Lorenzo in Firenze (Florence) where many members of the Medici family were interred. On 4 October 1881, at the close of the second International Geological Congress, a distinguished group of international delegates including the Canadian Thomas Streey Hunt, the American James Hall, the German Karl Alfred Zittel, and the Italian Giovanni Capellini (the President of the IGC) left Bologna and assembled at Steno's tomb in the Capella Stenoniana. Later a bust and a plaque in Latin were unveiled at the tomb. The plaque reads 'You behold here, traveller, the bust of Nicholas Steno, as it was set up by more than a thousand scientists from all over the world, as a memorial to him ... [he is] illustrious among geologists and anatomists.' Today his tomb is still venerated, and is the destination of many pilgrims. It is not unusual to find it covered with hastily scribbled notes on scraps of paper, or with letters, photographs or flowers, all left by pilgrims seeking indulgences or help.

He settled in Italy in 1665 and became fascinated with the Tuscan landscape, and with the fossils that he and others had found. By 1669 he published his great geological work, the *Prodromus* (or to give it its correct and fuller title *De solido intra solidum naturaliter contento*

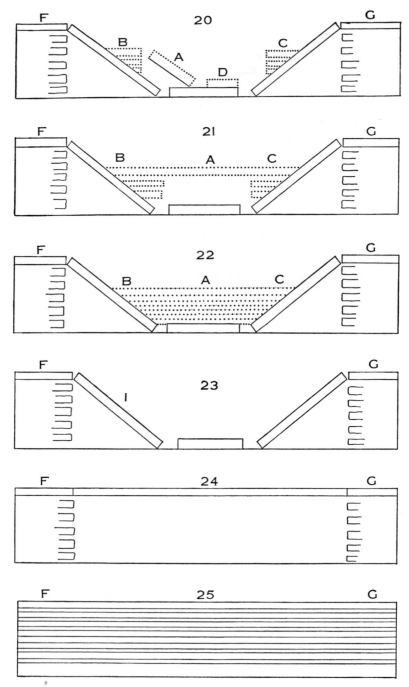

Figure 5.2   Steno's diagram 20–25 showing the stages of the development of the Tuscan landscape. It should be read from the bottom (oldest stage)

dissertationis prodromus), upon which most of his reputation is now based. Here he pulled together information on Tuscan rocks, fossils and landscape in order to work out the evolution of that landscape. He set aside the generally accepted assumption that these physical features had been formed at the Creation, as he comprehended their subsequent formation. In this book he made many geological observations that are now taken for granted by modern geologists: quartz crystals all displayed the same angle between faces; the major influence on the formation of topography was running water; like Leonard da Vinci (1452–1591) who based his conclusions on the nature of fossils on clear and painstaking observation of specimens found *in situ*, Steno recognised that fossils were the remains of once living organisms. He paid particular attention to the study of sharks' teeth which had until this time had been considered by many to be petrified tongues, and he also interpreted and illustrated the sedimentary record. However, he also made some pronouncements that now seem odd: coal and ash indicated the presence of former subterranean fires; and volcanoes occurred through the combustion of carbonaceous material at depth.

Steno explained that the Tuscan landscape had undergone transformation in six stages, which he illustrated in a clever and revealing diagram (Figure 5.2). It has been suggested by some commentators that

---

Fig. 5.2 (cont.)
to the top (youngest): (25) Strata are deposited in horizontal layers. (24) Erosion of an underground cavern has taken place in the central part of the diagram. (The strata beneath F and G on the left and right hand sides appear to be eroded, but this is not the case, and although Steno drew the diagram in this way, the strata should be taken to extend to the lateral margins of the diagrams in all cases 20 to 24). (23) The overlying strata have collapsed into the cavern and produced dipping strata. (22) Subsequently younger horizontal strata B, A, C (dotted) have been deposited on top of the older strata, and produced an unconformity. (21) Erosion of another cavern has occurred, but this time in the younger strata. (20) Further collapse of strata into this cavern has caused more dipping strata A, and later deposition of youngest rocks D has occurred. The dotted lines indicate argillaceous or sandy sequences that may be either unconsolidated or poorly consolidated while the solid lines represent predominantly lithified horizons. This diagram has been redrawn from the original (from Winter, *Prodromus* (1916), Plate 11).

Steno used these stages to represent each of the six days of the Creation week, and so may have been constraining an earthly chronology of 6,000 years. The six stages can be summarised as follows: in stage one, horizontal strata are deposited in the sea; stage two sees the regression or retreat of the sea leading to a drying out of the land and the formation of cavities beneath; in stage three, mountains and valleys are formed through the action of fires and water; stage four is marked by a return of marine conditions and the deposition of fossiliferous layers of sediment; in stage five another regression occurs and erosion of the rocks by river water takes place; finally in stage six the present-day landscape is formed by fire, water and collapse of the rock layers.

A colleague of mine in Dublin, the noted historian of geology Gordon Herries Davies, wondered why Steno's study of a small and seemingly insignificant area of limestone and their contained fossils in a part of Italy should be so influential. Why were Steno's findings important in the history of the development of geological thought? The answer lies in his observation that the nature of the Earth's history and development could be deciphered through an examination of the rock succession. In this he recognised unconformities, which represent a break in the stratigraphical record, and where often horizontal layers are found deposited on earlier tilted and eroded rocks. These are called angular unconformities, although horizontal unconformities do occur in the geological record, but the breaks in the sequences are naturally harder to spot. Although he illustrated such structures it is unclear whether Steno actually understood their significance as recording breaks in the geological succession. Steno did, though, recognise the importance of superposition, a principle more often associated with the English geologist William Smith (1769–1839), in which it is understood that in a sequence of rocks the underlying horizons are older than those above them, unless of course it can be demonstrated that the whole sequence has been overturned. Steno also recognised that horizons (layers of rock) found at either side of a valley that corresponded in terms of their position in the sequence and lithology were once connected by a lateral and

continuous horizon. This is known as his principle of lateral continuity. His third principle was that of original horizontality, which said
that any horizon recognised to be sedimentary in origin (and thus
deposited in water), but now found at any angle, would originally
have been horizontal.

Steno's work flew in the face of general geological understanding
and belief at the time, which still relied largely on biblical teachings.
His concepts moulded the geological thinking of others who followed
him. While his ground-breaking ideas do not tell us anything about the
actual age of the Earth, Steno did not set out to discover this fact. But
his work did allow geology to progress and thinkers to begin to understand the nature of the geological record. He showed, in short, that
geology does reveal a history.

'THE ENGLISH STENO'

Perhaps referring to Robert Hooke (1635–1703) as 'the English Steno'
is doing him a disservice; but he was ruminating on geological matters
at much the same time as Steno was reflecting on the hills of Tuscany,
and his conclusions are every bit as important as his Danish contemporary. Hooke has been unfortunate in that he has been largely overshadowed by his contemporaries Isaac Newton and the architect
Christopher Wren. Recently he has been in the ascendant – a natural
position given the celebrations of the three-hundredth anniversary of
his death. This was marked by the broadcasting in Britain of a television documentary and the publication of at least four books on his
life and work, one of which examined Hooke's hands-on role in the
phoenix-like reconstruction of London after the Great Fire – credit
that had hitherto been placed firmly at Wren's feet.

Robert Hooke was born in 1635, on 18 July, at Freshwater on the
Isle of Wight, close to the southernmost exposure of chalk in Britain.
His father was the local Church of England minister, but died young
when Robert was only thirteen. Hooke was then sent to London where
he served an apprenticeship under the artist Sir Peter Lely. Although
he found the smell of oil paint exacerbated his frequent headaches – he

was not a healthy young man – he took further lessons in art and sketching before going up to Oxford. There he came under the influence of Robert Boyle and became hooked on science. In 1662 he returned to London where he took up a non-stipendiary post as Curator of Experiments at the Royal Society, and in the following year was admitted as a Fellow, which brought him into close contact with all the academic thinkers in England at the time. His work there encompassed research into nearly every facet of science that existed. By 1664 he was being paid by the Royal Society but also secured a professorship at Gresham College in London. He was a prolific note-taker and kept numerous laboratory books that were festooned with scribbles, sketches of equipment designed by himself, notes, casual thoughts and sundry scientific results. Many of his observations on natural history were included in his *Micrographia, or Some physiological descriptions of minute bodies made by magnifying glasses, with observations and inquiries thereupon*, published in 1665. This was a treatise on the natural world as seen under the lens of a microscope, and is one of the earliest examples of this genre; such books became commonplace in the popular scientific market in the 1850s. Hooke was by no means an easy man, and was considered by many to be somewhat miserly, although he did leave a substantial fortune on his death. He clashed not infrequently with his colleagues in the Royal Society (including Newton, who he felt had plagiarised some of his data on the matter of the motion of the planets), but by 1677 he had been elected one of its secretaries and so felt that he held some power and influence. Hooke died in his rooms at Gresham College on 3 March 1703 and his effects were dispersed soon afterwards. Some found their way into the collections of the Royal Society, others simply vanished. He was given a 'Nobell funerall', at which his friends were offered fine wines by his executors and heirs, after which he was buried at St Helen's Church, Bishopsgate, in London.

Many of Hooke's geological observations were delivered in a series of lectures to the Royal Society in 1668, but not published until after his death, and are contained in the *Discourse of Earthquakes* which was edited by his friend Richard Waller in 1705. Other ideas, such as some on minerals, appeared in his *Micrographia*. Unlike Steno, Robert Hooke was not too concerned with framing his geological ideas within a religious context. While Steno was aware that minerals had particular shapes, Hooke's ideas, published four years earlier, were more complex and comprehensive. He realised that their external or crystallographic shape was controlled by an internal arrangement of matter, a concept he tried to promote and explain through the use of illustrations in which spheres (or 'globules' or 'bullets' as he called them), were packed in different arrangements to make up different outline shapes. Today we know that minerals have a defined and regular chemistry and atomic structure, which controls the external habit of the mineral species.

Having been brought up on the Isle of Wight, Hooke would have been familiar with the different arrays of fossils found in the various horizons that form fabulous coastal exposures at so many points around the island. He probably made collections of such specimens, and he was certainly conscious that they represented the remains of actual organisms. He had no time for the folklore that was attached to these past animals, and a fine suite of ammonites illustrates his work. These spectacular fossils were cephalopod molluscs, which were abundant in the Mesozoic oceans and were similar in morphology and life habit to the modern-day *Nautilus* that is found in the Indian Ocean. He also recognised that animals living today might not have been alive in the past, and conversely that some animals were present in the past but no longer present today: 'there have been many other species of Creature in former Ages, of which we can find none at present; and that 'tis not unlikely also but there may be divers [diverse] new kinds now, which have not been from the beginning.' He noted that past events could be recognised and dated by changes in the fossil record. This forecast the science of biostratigraphy, the dating of the geological record using fossils. He was also anticipating some of Charles Darwin's notions of over a hundred years later.

Hooke also propounded theories on strata, noting that water was the main agent of deposition but also of denudation, and he understood that air or wind could also carry sediment: loess, a type of wind-blown soil found in central Asia, is formed by such a mechanism. He also said that volcanoes produced their own sediment in the shape of ash that fell following an eruption and became compressed into rock through the pressure from overlying sediments, which is true; and that other methods of lithification were effected by sunlight which dried out surface sediments. While baked muds do form in this way, it is rare that they become consolidated and lithified unless they are buried by later sedimentation.

Hooke was also concerned with the generation of global movements that disturbed the strata. He suggested that these movements were activated by earthquakes which resulted in the continents moving around, which in turn resulted in climatic alteration from place to place. Such comments would not be out of place today. These earthquakes, he said, would destroy mountains and lead to the burial of surface rocks and the exposing of buried rocks. On older, uplifted eroded rocks he conceived younger sediments being laid on top. This was the first explanation of how an unconformity would be produced. We see in Hooke's *Discourse* a dynamic cyclical theory of the Earth, not fully developed, that found voice through a Scottish gentleman farmer some eighty years later. For this, and if only for this alone, Hooke deserves greater credit and acknowledgement: his geological ideas certainly influenced this Scottish gentleman, whose own writings laid the basis for the global theories now taken as read by geologists and by most of the general public. Thanks to the long-standing efforts of a few historians of geology, most notably Ellen Tan Drake, and the celebrations that marked the tercentenary of his death, Hooke has at last joined the pantheon of seventeenth-century scientific worthies.

LOCAL THOUGHTS ON LOCAL STRATA

While today Hooke's and Steno's ideas in stratigraphy are held as being the first in this subdiscipline of geology, it is difficult to appreciate just how great their influence was. Certainly the illustration of unconformities by Steno was important, and these geological features were later shown by James Hutton (1726–1797) to be of the utmost value in unravelling past geological histories (see Chapter 6). From the 1700s many people across Europe, whether aristocrats, gentlemen of science, or lowly mineral prospectors, shared a common interest in geology, and many of these would have been familiar with at least one of the various language editions of Steno's *Prodromus*. A number of these readers realised that rocks were layered, or stratified, and it was therefore natural for attempts to be made to classify these successions, explain the differentiation of rock types and describe their composition and structure.

One of the earliest definitions of stratification was that of the Reverend John Michell (1724–1793), according to the later Scottish geologist Archibald Geikie who quoted it verbatim:

> The earth is not comprised of heaps of matter casually thrown together, but of regular and uniform strata. These strata, though they frequently do not exceed a few feet, or perhaps a few inches in thickness, yet often extend in length and breadth for many miles, and this without varying their thickness considerably.

Michell, like many university geologists then and later, was appointed to a chair in geology (in his case the Woodwardian Professor of Geology at Cambridge) on the strength of little research in the field. An English country rector, he wrote an important discourse on earthquakes, which he based on observations of the catastrophic Lisbon earthquake of 1755. This paper, which was read to the Royal Society of London in 1760, is important in that Michell suggested for the first time that movement waves accompanied earthquakes. These radiating waves, if mapped, he argued, could point to the centre of the earthquake. It was nearly a century later that the Irish engineer Robert Mallet (1810–1881) proposed the term 'epicentre' for the origin of the earthquake. At the time of Michell's paper it was thought that earthquakes were caused by huge volumes of water vapour produced when water

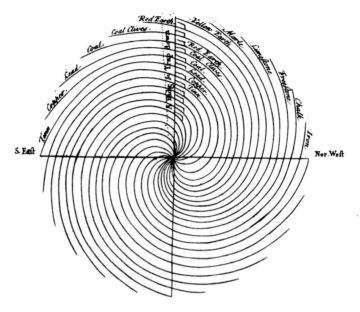

Figure 5.3   John Strachey's 1725 cross-section through the Earth showing dipping strata caused by its rotation (from B. D. Webby, *Proceedings of the Geologists' Association* **80** (1969), 91–97, Plate 5).

quenched subterranean fires thought to be located beneath the Earth's crust. This head of steam exerted pressure on overlying strata which triggered the shockwaves.

Others attempted to explain why strata dipped. Not all sedimentary beds are horizontal; indeed, it is common to find inclined strata, which we now attribute to the effects of tectonic movement of the Earth's crust. In the 1700s no one knew that earth movements could occur on a scale large enough to move lithospheric plates or continents. Inclined strata were explained away by invoking several ideas: John Ray noted that coal miners thought that beds dipped towards the centre of the Earth; and the English mining surveyor John Strachey (1671–1743) suggested in 1725, in a paper published in the *Philosophical Transactions of the Royal Society*, that strata became formed and separated from each other (and became inclined) because of the rotational effect of the Earth (Figure 5.3), an idea modified from that published by William Stukeley (1687–1765) a year earlier. Strachey imagined

> the Mass of the Terraqueous Globe to consist of ... perhaps, of ten
> thousand other different Minerals, all originally, whilst in a soft and
> fluid State, tending towards the Centre. It must mechanically ...
> follow, by the continual Revolution of the crude Mass from West to
> East [...] like the winding up of a jack, or rolling up the Leaves of a
> Paper-Book, that every one of these Strata ... appear to the Day
> [Earth's surface, with the] lightest to be uppermost ....

As John Fuller has pointed out, Strachey was the first to use stratigraphical cross-sections to illustrate the disposition of the geological
structure beneath the surface. Use of this graphic innovation later
became widespread and today is one of the first concepts taught to
undergraduate geological students setting out to map for themselves.

EARLY EUROPEAN DEVELOPMENTS TOWARDS A GLOBAL
STRATIGRAPHIC FRAMEWORK

While most eighteenth-century ideas in stratigraphy such as those of
Ray and Strachey were confined to explanation of local phenonema, a
number of authors expanded local information into a global stratigraphy. At this time the rapid rise in geological exploration and localised
mapping was in response to the need for basic raw materials such as iron
ore and coal, particularly at times when various nations were engaged in
hostilities with each other. The German or Prussian states were largely
fragmentary, but under threat at times from France in the west and from
the Austro-Hungarian Empire to the southeast, which itself was continually skirmishing with the northeastern Italian states of Venice and
the Veneto. In the Germanic states various mining academies were
active across the country; that at Freiberg, presided over by the eminent
Abraham Gottlob Werner (1749–1817), was perhaps the best known.
This drew students from far beyond the national boundaries, with
English, Scottish, Irish, French and American students known to have
enrolled. Equally in Italy, scientists in institutions and academies

were particularly interested in the origin of mountains and the nature of volcanoes. This is not surprising given the central axis of the Apennines and the active volcanic centres in Naples, in the Aeolian Islands of Stromboli and Vulcano, and in Sicily.

On continental Europe a large number of individuals attempted to construct a general stratigraphical framework. Among these earliest attempts, the most noteworthy were those of Johann Gottlob Lehmann (1719–1767), Giovanni Arduino (1714–1795), Torbern Olof Bergman (1735–1784) and Werner. The writings of others were further from the mark – for example, Benoît de Maillet in his book *Telliamed* suggested that the oceans covered the whole globe, and that they were responsible for the deposition of mountain ranges and the sculpturing of the Earth's surface. However, as we have seen, he ventured further to propagate the idea that the oceans were progressively shrinking, and had been over a period of two billion years. Two billion years – such a timeframe was unimaginable, and the reaction from the church authorities soon caused de Maillet's radical theories to founder. One must remember that these ideas were formulated at a time when the biblical thinking on the origin of the Earth was still very much in fashion, and that the church still had great powers of persuasion against such anti-biblical ideas, affecting both the readers of these works and sometimes even their authors.

Lehmann was a teacher, mining engineer and surveyor who was familiar with the geology of his native Prussia. He published a number of books which advanced some interesting ideas concerning the origins and structure of mountains; he argued that volcanoes and earthquakes were important processes in moulding the surface topography of the Earth; and he thought that the crust was made up of a series of as many as thirty layers laid down under water. At the same time, he promoted some ideas that would nowadays be greeted with laughter. One of his celebrated, and oft-quoted, suggestions was that gold was produced by the Sun, as it was more frequently found in areas of hot climate rather than in colder latitudes. But Lehmann's important contribution to stratigraphy was in description of a tripartite division of mountains: *Primitive* – hard, comprising non-fossiliferous crystalline rocks; *Flötzgebirge* – composed of stratified, water-laid, often fossiliferous rocks; and *Alluvial* – younger mountains that comprised unconsolidated sediments or recent volcanic materials.

Giovanni Arduino was an Italian who worked at different periods in his life as an inspector of mines or as an agricultural advisor. Born in the village of Caprino, near the beautiful city of Verona where he was also educated, Arduino became fascinated with the natural landscape around him, and began to read the geological writings of Woodward and Burnett amongst others. However, according to Frank Dawson Adams, he did not learn much geology from these authors, and so began to examine the local geology for himself in an attempt to learn more about the subject. He really could not have chosen a better field area: he had the rocks of the neighbouring Alps and the sediments of the flat plain of the River Po to inspire him (Figure 5.4). No doubt he was also familiar with the volcanic regions of the area that was to become known as Kingdom of the Two Sicilies – Naples and the Island of Sicily. From the age of eighteen he was employed in various mining regions as a mining inspector. Although he did not publish extensively, he did correspond widely: his letters hold much valuable information on geology, and give an insight into the development of his theories and of his understanding of earth processes and the origin and constitution of mountains. Arduino's major contribution to stratigraphy was in his *Due lettere sopra varie osservazioni naturali* published in 1760, in which he presented a tripartite classification of rocks into 'ordini' – *Primary*, *Secondary* and *Tertiary*. The *Primary* sequences were the basal cores of mountains, which were largely altered metamorphic successions that were often folded and criss-crossed with veins of mineral-bearing white quartz. Such veins may contain considerable proportions of base metals such as lead and zinc, and minor percentages of the precious metals gold and silver. The *Secondary* sequence comprised fossiliferous limestones and marls, as well as marbles of various tints and hues, while the *Tertiary* were younger limestones, marls and siltstones formed into the lower mountains, or

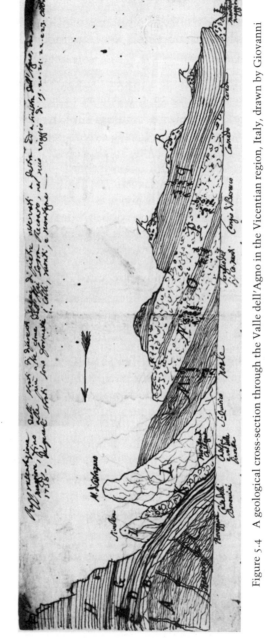

Figure 5.4   A geological cross-section through the Valle dell'Agno in the Vicentian region, Italy, drawn by Giovanni Arduino (1714–1795) in 1758 (from Ezio Vaccari, *Giovanni Arduino (1714–1795)* (Florence: Olschki, 1993), Figure 9).

Pre-Alps, that are plastered up against the Alpine mountains proper. Some of these younger sediments showed evidence of being reworked by geological agents from the older Secondary series. Arduino also erected a fourth 'ordinus', the *Quaternary*, for the more recent volcanic products such as lava, volcanic ash and tuff. This fourth category also included the unconsolidated alluvial derived from the three older 'ordini'. Undoubtedly this category also included other similar deposits often found blanketing the landscape, glacial origins of which were not recognised until nearly a century later. Arduino also recognised that many of the rocks he observed could not be attributed to the Flood, but were the products of many different periods which had undergone uplift, folding and alteration. This was an important and early observation of the dynamic nature of the Earth in which geological features were not the results of one event but the culmination of many over some long period. Arduino, unlike Lehmann and many others at this time, was able to disregard the time limitation effected by the biblical Creation, and he appreciated that the Earth's time-span was far longer, although he was unable to tell how great.

Bergman, who had studied under Carl von Linné (Linnaeus) at the University of Uppsala, became Professor of Chemistry at the same university in 1767 and wrote on a wide diversity of subjects. He was a noted chemist, mineralogist, astronomer, and in the context of this present discussion, a geologist. He is primarily recalled today for having coined the term 'ammonia' for the foul-smelling gas nitrogen hydride. In Roman times one particular ammonium chlorite deposit in Libya was known as the salt of Amun or *sal ammoniacus* because it occurred around an oasis close to the Temple of Jupiter Amun. These deposits were produced by visiting camels as they defecated and urinated. Bergman latched onto this for his name, the etymological origin of which could mistakenly be taken to be derived from Amun the Egyptian god of fertility. Ammonia is a great fertiliser, but this connection with Amun is sadly coincidental. He does however lend his name to the wonderful group of extinct cephalopods, the ammonites, which were closely allied to the modern-day *Nautilus*. These fossils were originally thought to resemble rams' horns – horns which Amun was often depicted wearing as a head-dress.

The year before his professorial appointment Bergman published a weighty volume titled *Physical Description of the Earth*. This volume was revised and expanded into two volumes published in 1773 and went into several foreign language editions. Bergman's geological opinions and observations were widely disseminated and may even have influenced the thinking of Werner. Bergman argued that all rocks were precipitated from water, but that this took place episodically and locally. He recognised four major divisions of strata that were arranged in distinct layers in the Earth: the *Uråldrige* or primitive rocks, by which he meant principally crystalline igneous rocks such as granite, or metamorphic gneisses, formed the innermost layer and were found in the cores of mountains. The *Flolågrige* or bedded rocks, which made up the second layer, comprised sandstones, coals and limestones, which Bergmann regarded as having been formed from suspended materials that settled out of the waters. The sediments were derived by the erosion of the rocks of the *Uråldrige*. The third layer he termed the *Hopvråkta*, which broadly means 'swept together', and these were the largely unconsolidated sediments found on the Earth's surface. The final layer, the *Vulkaner*, were volcanic products produced by melting in the deep fires thought to be found in the interior of the globe, and subsequently erupted on to the surface where they cooled and crystallised.

Abraham Gottlob Werner, our fourth eighteenth-century definer of a stratigraphical framework, was probably the most widely known, as we shall see later, partly because of his role in the debate between the so-called Neptunists and the Vulcanists. Werner was born in Wehrau on 25 September 1749 and died at the age of 67 in Dresden on 30 June 1817.

Werner joined the staff of the Bergakademie (mining academy) in Freiberg ten years after its establishment by the Elector of Saxony in 1765, and to his professorship of mineralogy he later added the title of 'Councillor for Mines in Saxony'. At the Bergakademie students followed a three- or four-year course in chemistry, mining methods, mineralogy and mathematics. Werner taught amongst other courses geognosy – defined as the science or theory of the formation of the Earth; petrography – the study of rocks; and mineralogy – the study of minerals, particularly those applied to mining. He was a meticulous teacher both in the laboratory and in the field and instilled in his students the value of making observations and recording their findings either in notebooks or as geological maps and plans.

Over the years Werner examined large tracts of Saxony, its landscape and its mines, and in doing so acquired an encyclopaedic knowledge of the geological make-up of the region. It was natural that in his ordered mind he should start to formulate a classification of the rocks with which he had now become so familiar. This classification probably underwent some evolution through time, but it was most famously expounded and laid down in his short 28-page publication *Kurze Klassifikation und Beschreibung der verschiedenen Gebürgsarten* which he completed in 1785 and published in Prague the following year. Alexander Ospovat, the foremost student of Werner in the past fifty years, said of it: '[it] guided geological observations and formed the basis for rock classifications ... from 1786 until about 1825. It established petrography as an independent branch of the geological sciences and made the doctrine of geological succession a cardinal principle of earth history.' Broadly speaking Werner devised a four-fold classification of the rock succession: (1) *Urgebirge*: 'primitive' rocks (the oldest and lowermost) such as granite, gneiss, marble, quartzite and basalt in which fossils were not found; (2) *Übergangsgebirge*: 'transitional' rocks such as clastic sedimentary rocks and limestones with frequent fossils; (3) *Flötzgebirge*: 'Floetz' rocks which were well-bedded limestones, shales, coals, clays and sandstones with frequent fossils; and finally (4) *Aufgeschwemmte Gebirge*: the overlying (and youngest) 'alluvial' rocks which comprised unconsolidated gravels, sands, soils and peat. In these he recognised many reworked fossils from the underlying Floetz. He also suggested that volcanic rocks (such as tuff, lava, ash and pumice) could occur at any level in the scheme – note that he failed to recognise that basalt was a product of volcanic activity. Werner imagined that these rocks were deposited during a global flood event, and the primitive rocks formed a basal blanket over the whole globe, on which the sediments of the later periods were subsequently deposited. As the rocks built up in the oceans some successions broke through the water and became the continental masses. Werner's *Kurze Klassifikation* is important as it was an attempt to provide an integrated classification of all the rock types known on the Earth's crust, and it is, surprisingly, the only such scheme to come from Werner himself.

As Ospovat has noted, Werner's scheme was widely 'adopted, adapted and sometimes just copied.' Werner's influence in geological matters was widespread thanks to his teaching role in Freiburg. Many of his students became notable – no, this is too weak a description – geological giants in their own right. They included Friedrich Wilhelm Heinrich Alexander von Humboldt (1769–1859), the traveller and author of the geological and geographical classic *Kosmos*, Christian Leopold von Buch (1774–1853), who formulated a theory on uplift and the formation of volcanic craters and who edited the first geological map of Germany, and Ernst Friedrich von Schlotheim, who had a glittering palaeontological career in Russia. In Britain his greatest disciple was Robert Jameson (1774–1854) who returned to Edinburgh to the Chair of Natural History in 1804 and soon afterwards established the Wernerian Natural History Society in that city.

While Werner did not discuss the actual age of the Earth, he noted in 1786 'the enormously great time spans which perhaps far exceed our imagination'. This comment remained unpublished for nearly two hundred years until 1971 when it appeared in a volume that contained a classification of various rocks, a manuscript classification brought to light and edited by Ospovat. Werner verbalised the great difficulty that many educated people faced at the time: from the evidence of the rocks geological time was vast, but still biblical and religious dogma played its part in keeping at bay the widespread public proliferation of such thoughts.

Of these four men, only Arduino steered clear of biblical explanations for the formation of the various lithologies that he described. The classifications erected by the other three were dependent on the largely accepted premise that the water thought to have engulfed the world during the period of Noah was the major geological process that acted during the formation of the Earth's crustal and lower rocks. It is perhaps surprising in hindsight that they did not stop to consider that the story of the Flood was related by scribes writing about the events that affected a small geographical region – the present day Middle East, Iraq and Iran – and that the resultant geological effects (if any) of such events could not be stretched to cover western Europe and beyond. No doubt the authors of the Bible saw the effects of earthquakes which would have occurred moderately frequently along the zone of crustal weakness that runs from modern-day Turkey, south through Syria, through the Dead Sea zone and into the Red Sea. Some major catastrophic happenings related in the Bible, such as the destruction of Sodom and Gormorrah, the tumbling down of the walls of Jericho, and the Flood, can all be explained rationally by attributing them to the effects of an earthquake. Through the passage of time, such events would have become embellished and explained in a way that the listeners could rationalise within the compass of their own experience. It was far easier to invoke God, rather than understand the complex causes and patterns of earthquakes.

The conceptual framework for the understanding of geological history, as read through the rock successions, was now in place, and geologists had a skeletal stratigraphical manual containing blank pages waiting to be filled with additional lithological, biostratigraphical and chronostratigraphical data.

6   An infinite and cyclical Earth and religious orthodoxy

In the 1990s, when the Geological Gallery at the American Museum of Natural History in New York was being redesigned, a series of moulds was made of classic geological sites and these were cast in fibreglass and installed in the geology exhibition. While there one can gaze at a piece of Scotland: Siccar Point on the Berwickshire coast, to be precise, which is one of the sites in the world that historians of geology would wish to visit to pay homage to the early geological fathers. Recently I was in New York and made my way to the gallery. I have to confess that I have never visited the real Siccar Point. Nevertheless its impact on me was palpable – I had to sit down in the gallery, and I gazed at the structure, identifying the almost upright beds overlain unconformably by almost horizontal beds tilted at a low angle. I realised that perhaps I was experiencing a similar sensation to that of the great gentleman geologist James Hutton (1726–1797) when he first saw the actual site over two hundred years earlier.

THE GENESIS OF A NEW THEORY OF THE EARTH

Sitting in the parlour of his Edinburgh house at St John's Hill on the afternoon of 7 March 1785, the regal-looking gentleman would have been somewhat apprehensive. It was nearly teatime but I suspect he would not have wanted to eat much; he had more pressing matters on his mind. That evening an outline of his ideas on the Earth and its geological history was to be delivered to the Royal Society of Edinburgh.

James Hutton (Figure 6.1) was born on 3 June 1726 in Edinburgh. His father William was a merchant and one-time City Treasurer, but he died, leaving a widow Sarah (née Balfour) and four children: James and his three sisters. In 1743 James became a clerk in a local firm but

Figure 6.1   James Hutton
(1726–1797) (from John Kay,
A Series of Original Portraits,
vol. 1 (1838), in Davies, *The
Earth in Decay* (1969), Plate 1).

found this occupation unsuited to his temperament, and so in 1744 matriculated at the University of Edinburgh and began his studies in the humanities while at the same time studying medicine under the tutelage of Dr George Young. However, like many students he was unsure of his calling and began to show an interest in chemistry. It was at this time that Hutton, who never married, fathered a son. Following a period spent on the Continent – in Paris, where he studied anatomy and chemistry, and Leiden, where he gained the degree of Doctor of Medicine – Hutton headed back to Scotland in 1750 and decided to throw his energies into farming, unsurprising as he had inherited from his father some land near Slighhouses in Berwickshire 50 miles east of Edinburgh. A progressive man, he decided that in order to make the most of this career choice he should study the newest methods in agriculture, and so spent some time in Norfolk, in England, where he learnt a great deal about husbandry, in particular from his landlord, the farmer John Dybold. He also travelled in Flanders for the same reasons, but began to take a serious interest in geology and mineralogy. From 1754 until 1767 he farmed his land. On leaving Norfolk he bought a plough and employed a ploughman and returned to Scotland where he introduced novel agricultural methods to his district. He later wrote a long treatise on these methods and a 1,045-page manuscript on his agricultural philosophy, *Elements of Agriculture*, which interestingly contained some opinions on biological evolution of organisms, but this unfortunately remains unpublished. Although some distance from Edinburgh, he maintained his contacts with the intellectual circle, and it was to this group of friends that he turned when he gave up farming for good and moved permanently into Edinburgh in 1768. There he joined the Philosophical Society and later established the Oyster Club with Joseph Black (1728–1799), who discovered carbon dioxide, and Adam Smith (1723–1790), the economist whose will he executed. The club became a meeting place which saw informal weekly gatherings of like-minded people such as Sir James Hall (1761–1832) of Dunglass, John Playfair (1748–1819) and others. Playfair in his *Biographical Account of the Late Dr James Hutton*, published shortly after Hutton's death, notes that Hutton ceased farming because once he had established good farming methods on his properties, as 'the management of it [his farms] became more easy, it grew less interesting'.

Hutton's interest in geology seems to have beeen sparked in about 1752. What might have caused this? This is difficult to answer, but he may well have been introduced to some geological works by members of the Scottish Enlightenment. There is very clear evidence from Hutton's own writings on the subject that he had read Hooke's *Dissertation on Earthquakes* and *Micrographia*, although the Scottish savant never acknowledged this as such. He may well have read Steno and Burnet and other early treatises, including those by his fellow Scot George Hoggart Toulmin (1754–1817) who argued in four volumes published between 1780 and 1789 that the world was eternal. A number of historians of geology have rejected Hutton's reliance on Toulmin's work. Nevertheless, it is now generally accepted that some of Hutton's ideas were not simply a 'road to Damascus' spontaneous revelation, but must have been a conscious or subconscious reworking and reformulation of earlier ideas, largely those of Hooke, to which he added considerable insights and observations of his own, formed while engaged in field work and travel.

As well as his travels in England in the 1750s and on the Continent, Hutton went on a long tour of the north of Scotland in 1764 in the company of his close friend George Clerk, later Clerk Maxwell (1713–1784). In 1774 he toured England and Wales when a visit to the salt mines in Cheshire with James Watt (1736–1819) made a considerable impression. Hutton was interested in the quartzose gravel which underlies much of Birmingham, and visited Wales to try to discover the source of this material. In doing this, he demonstrated his early appreciation of the role of denudation in the formation of later sediments, and thus later geological horizons. He was unsuccessful in finding the source of the sediment until he returned to Birmingham where he found a suitable lithology locally. In 1777 Hutton published a short paper that contrasted and compared the coal successions in Scotland with those found in England, and it was largely thanks to this work that the Scottish coal producers were exempt from paying government duty so long as the fuel was transported to its destination by sea. Hutton also examined the structure of Arthur's Seat in Edinburgh and published a paper on the subject in the *Transactions of the Royal Society of Edinburgh*. It seems that he read several papers to the Royal Society but failed to publish them – he tended to be reluctant to go to press with new information, and much of his work was published thanks to generous encouragement from his friends, or as a result of opinions voiced by those who disagreed with his theories.

HUTTON'S THEORY OF 1785

Following about thirty years of study and occasional travel, Hutton was ready to launch his theories on to the world stage. On 7 March 1785, the date that the first part of his work was read to the assembled Fellows of the Physical Class of the Royal Society of Edinburgh,

ABSTRACT

OF A

DISSERTATION

READ IN THE

ROYAL SOCIETY OF EDINBURGH,

UPON THE

SEVENTH OF MARCH, AND FOURTH OF APRIL,

M,DCC,LXXXV,

CONCERNING THE

SYSTEM OF THE EARTH,

ITS DURATION, AND STABILITY.

Figure 6.2 Title page of Hutton's 1785 abstract.

Hutton was not in the chamber, and the paper was read by Joseph Black. Quite why Hutton was absent at this critical time is not known, but perhaps he could not face any hostility that his ideas might have engendered in the assembly. On 4 April following, Hutton was present, and he read the second half of his paper. The anonymous thirty-page abstract (Figure 6.2) of this paper was published later that year and an expanded version was published by the society three years later in 1788. Copies of Hutton's 1788 abstract are exceedingly rare; it was only recognised as being Hutton's work in 1947, when this was proved by the historian of geology Victor Eyles. Gordon Craig, a noted scholar of Huttoniana, in his introduction to a facsimile edition of the abstract produced in 1997 to mark the bicentenary celebrations of Hutton's death, remarked that he knew of only seven copies in public circulation.

Hutton began his 1785 abstract by outlining the aims of his work:

The purpose of the Dissertation is to form some estimate with
regard to the time the globe of this Earth has existed, as a world
maintaining plants and animals; to reason with regard to the
changes which the earth has undergone; and to see how far an end or
termination to this system of things may be perceived, from the
consideration of that which has already come to pass.

He went on to note on page 5:

That the land on which we rest is not simple and original, but
that it is a composition, and had been formed by the operation of
second causes.

He realised that the sediments had to have become consolidated, and
he suggests that the mechanisms by which this could have happened
was through the cementation of sediment particles by cements precipitated from sea water, or by the action of heat, which caused the
unconsolidated sediments to fuse together. He argued that the sedimentary rocks would have had to have been uplifted from their place
of deposition in the sea, to a point where they were on dry land above
sea level. How did he explain the second problem, that of uplift? He
invoked the same heat of fusion which he said 'might be capable of
producing an expansive force, sufficient for elevating the land, from
the bottom of the ocean, to the place it now occupies above the surface
of the sea.' This land mass, he pointed out, consisted of irregular
twisted, folded and fractured rock which had been produced by the
subterranean heat. He went on to discuss the disposition of 'veins'
which he had observed to cross-cut pre-existing rocks, and which he
considered to be the products of volcanic melting. Importantly he
identified these veins as being basaltic (they were sills and dykes in
modern terminology) and he distinguished them from modern volcanic lavas. Modestly Hutton contended that:

There is nothing visionary in this theory, [it] appears from its having
been rationally deduced from natural events, from things which

have already happened; things which have left, in the particular
constitutions of bodies, proper traces of the manner of their production; and things which may be examined with all the accuracy,
or reasoned upon with all the light, that science can afford.

He recognised the great antiquity of the Earth and the cyclical nature
of geological processes and said how to find such evidence of deep
time:

we are led to conclude, that, if this part of the earth which we now
inhabit had been produced, in the course of time, from the materials
of a former earth, we should in the examination of our land, find
data from which to reason, with regard to the nature of that world.

He came up with no actual figure for the length of time that the Earth
had existed but said that by examining the rate of erosion of the rocks
at the surface and the deposition of the products in the oceans 'we
might discover the actual duration of a former earth'. That duration
was 'an indefinite space of time'.

Hutton expanded the abstract and it appeared in paper form in
1788 in the first volume of the *Transactions of the Royal Society of
Edinburgh*. Here was a new theory of the Earth that was quite different
from those that had been propounded before. The Earth was of indefinite
age, was ancient and its dynamic nature was cyclical. He concluded:

For having, in the natural history of this earth, seen a succession of
worlds, we may from this conclude that there is a system in nature;
in like manner as, from seeing revolutions of the planets, it is
concluded, that there is a system by which they are intended to
continue those revolutions. But if the succession of worlds is established in the system of nature, it is vain to look for anything higher in
the origin of the earth. The result, therefore, of our present enquiry is,
that we find no vestige of a beginning, – no prospect of an end.

Hutton's work was of huge significance as it allowed enlightened men
of science and learning to shake off religious chronologies and dogma.

BACKLASH

It was not surprising that there should be a backlash to James Hutton's *Abstract*, as it was probably circulated to various learned and scientific circles, such as the Royal Society in London, the Royal Irish Academy in Dublin and the various academies in major centres of science such as Uppsala, Stockholm, Berlin and Paris, to name but a few. Opinion was divided.

One debate centred on the origin and nature of basalt and other rocks. One body of thought, collectively the 'Neptunists', or to use a contemporary term 'Watermen', felt that these rocks were precipitates from water – a view held by Abraham Werner and most of his students in the Freiberg School of Mines. They regarded lava as being a product of volcanoes, but basalt as being lithologically quite distinct from it. They held that where volcanoes occurred in basaltic regions, they had formed after the basalt had been deposited.

The other group, variably known as 'Vulcanists' or 'Plutonists' or 'Firemen', according to the rock types in which they were interested, argued that the rocks had once been heated and were the products of volcanoes or other igneous mechanisms. The Vulcanist theory was proposed by the Frenchman Nicholas Desmarest (1725–1815), who, having observed the extinct volcanoes in central France, advocated in 1771 that basalt was the product of volcanic activity. Hutton proposed a plutonic origin for some igneous rocks, and noted that these were emplaced from a hot magmatic source from below.

In the British Isles the most vociferous attacks on Hutton and on Desmarest emanated from Richard Kirwan (1733–1812), an eccentric Dublin-based gentleman, who continued to advance the ideas pertaining to the biblical Flood. Kirwan was outraged and went into print defending the Neptunian theory in papers published from 1793 in the *Transactions of the Royal Irish Academy* and in his important book *Geological Essays*, published in 1799, two years after Hutton's death. In publishing his earlier papers he spurred Hutton into producing a full version of his theory; for this we have to thank Kirwan. In time a great deal of what Hutton and indeed Desmarest wrote was accepted by the global geological community, but not without a fight from the Neptunists.

Richard Kirwan was born on 1 August 1733 in Galway, Ireland, into a prominent Catholic land-owning family, and went on to become a noted chemist and geologist. He served as President of the Academy from 1799 until 1812. His book *Elements of Mineralogy*, published in 1784 (a second edition appeared in 1794), was the first English-language text on the subject, and elevated him to the ranks of the most influential mineralogists of his day.

He showed a remarkable talent for learning from an early age, and studied in Poitiers in France where he began to purchase books on chemistry, much to the consternation of his mother. He later moved to St Omer in order to study for the priesthood. There he excelled at classics and was appointed Professor of Humanities. In 1755, when he was twenty-two years old, his elder brother was killed in a duel, apparently with a porter of the House of Commons, and he succeeded to the family estates and an income of £4,000 a year. In 1761 he moved to London where he was called to the Bar, but he abandoned law in 1868 in preference for a life engaged in scientific study and endeavour. He returned to Galway in 1772 and spent the next nine years learning Greek and other European languages, and assembling a fine library. In 1777 he returned to London where he soon became involved in several societies, such as the Royal Society of which he became a Fellow in 1780, and the less influential group, the Chapter Coffee House Society. He was unfortunate in that his library was stolen by privateers while it was in transit on the high seas. Eventually the library made its way to the Salem Athenaeum in Salem, Massachusetts, where it can still be consulted. In London Kirwan carried out much of his work on chemistry, and for it he was awarded the Copley medal of the Royal Society. In 1787 his most important work in chemistry, *An Essay on Phlogiston and the Constitution of Acids*, was published. In this work he advanced the theory that phlogiston was a constituent of all combustible substances which, when burnt, lost phlogiston, broadly equivalent to a loss of oxygen. This theory was later challenged by the French chemist Lavoisier and Kirwan's fellow countryman William Higgins, who showed that there was an increase in weight on the combustion of metals, and that oxygen was absorbed in the process.

In later life Kirwan became a noted eccentric who detested flies so much that he paid his servants for each corpse presented to him. He disliked replying to correspondence and had his door knocker removed at 7 o'clock each evening so as to prevent further visitors gaining entry. He died in Dublin while engaged in the common practice for the time of starving a cold. His funeral was a glittering occasion attended by 900 city worthies. Today Kirwan lies in an unmarked grave in the graveyard of St George's Church, Hill Street, Dublin, which is now a tarmacadamed patch of waste ground frequented by football-playing youths.

Some of Kirwan's objections to Hutton's ideas, which are some-what difficult to glean from the contorted logic and language of his papers, can be summarised briefly. He was particularly concerned with the lack of evidence presented by Hutton, and said that the theory was at variance with geological knowledge at that time. Jean-André de Luc (1727–1817), who coined the term 'geology', similarly criticised Hutton for not having carried out enough field work and for having spent too much time indoors – grossly unfair, given the field work Hutton had undertaken before 1785. Kirwan himself believed that Earth history and structure could be explained with reference to the Bible, and subscribed to the theory that all rocks were precipitated from some primordial fluid, and that these were later eroded and shaped by the waters of the Flood. De Luc's own theory of the Earth treated the Flood as being the energy source for a time of global change. Kirwan regarded Galway Bay as being the site of an ancient granite body which was removed by these waters. He suggested that the general and specific attractiveness of particles caused differentiation in the early Earth. He believed that coal was a product of breakdown of granite and basalt to yield bitumen which then aggregated into dis-crete layers. Kirwan said that basalt rested on various rock types

including coal, limestone, gneiss and granite in which no evidence of
the effect of heat could be seen. Basalt contains zeolites, rich in water,
and calcite, which contains 'fixed air', and so could not have been
heated. He also objected to the findings of Sir James Hall who had
taken basalt and melted it, then allowed it to recrystallise, in order to
show the similarity of the recrystallised with the original, thus proving that basalt was magmatic (igneous) in origin. The recrystallised
basalt, like the original, contained minute air bubbles. Kirwan had
previously argued that such bubbles were characteristic of basalt
deposited in water, an argument that failed to find favour with the
Scottish petrologists. As Peter Wyllie of the California Institute of
Technology has written, these experiments earned Hall the later
title of 'Father of Experimental Petrology'.

IDEAS ON THE ORIGIN OF THE GIANT'S CAUSEWAY
The Giant's Causeway (Figure 6.3) in northeast Ireland is one of the
geological wonders of the world. In 1740 the Dublin Society's promotion of art was responsible for bringing it to the attention of many
people in Europe. That year the Society offered a £25 premium for art
and one of the entries was from a Susanna Drury, a Dublin artist. She
submitted several excellent canvasses of views of the Giant's
Causeway in County Antrim, which she painted over a period of
three months. Miss Drury was awarded the premium and subsequently her images were engraved in London by François Vivarès. It
was these prints that were distributed across Europe and that initiated
a steady stream of visitors to the Causeway coast. The Lord Bishop,
Augustus Hervey (1730–1803), Bishop of Derry and fourth Earl of
Bristol, apparently installed steps down to the site. Visitors included
John Wesley in 1778, John Whitehurst in 1783, Abraham Mills in
1787–88, Humphry Davy in 1806 and Jean-François Berger in 1811. It
is not outrageous to suggest that Susanna Drury is the most important
person in the history of Irish geology.

Nicholas Desmarest never viewed the Giant's Causeway, but on
seeing Vivarès' engravings he declared that the basalts that formed the

Figure 6.3   The Giant's Causeway, Co. Antrim, Ireland, viewed from the east (from William Hamilton Drummond, *The Giants' Causeway, a Poem* (1811)).

region were volcanic in origin. Whitehurst suggested that the columnar basalts formed as the molten basalt cooled. In the same year the Reverend William Hamilton (1755–1797) published an important and influential memoir entitled *Letters Concerning the Northern Coast of the County of Antrim* which advanced Desmarest's theory of the igneous origin of the Causeway rocks and in which he accurately described the geology of the coast. He was one of the founders of the Royal Irish Academy in 1785 and in 1790 he became rector and a local magistrate of an isolated Donegal parish. He was murdered in 1797 after local unrest, which would culminate in the 1798 rebellion. He left a widow and nine children who were granted monies by Parliament.

While Kirwan was loud in his rejection of Hutton and Desmarest's views on the origin of basalt, the Reverend William Richardson (1740–1820) was thunderous, and perhaps with good reason, for he actually spent some of the year living at Portrush, on the Antrim coast close to the Causeway. William Richardson had been elected a Fellow of Trinity College Dublin in 1766, took holy orders and eventually resigned in September 1783 on taking up the living of the parish of Clonfeacle in County Tyrone. Through his membership of the Royal Irish Academy he would have become familiar with the arguments surrounding the geology of the Giant's Causeway.

Richardson examined the ground for himself, and published a number of papers between 1802 and 1812. He was clearly upset at others discussing the geology of an area which they had visited briefly or not at all. He believed he had the authority, as a long-term resident, to make his views known. While he claimed not to subscribe to any theory or belong to any faction, he held (perhaps unwittingly) to the Neptunist theory for the origin of the basalt. He rejected the alternative theories on a number of grounds:

1. There was no evidence of a volcanic mountain or cone in Antrim;
2. Plants are found developed on lava flows but not in between the basalt layers;
3. The constituents of basalt and lava were different;
4. The layers of basalt were horizontal and regular in thickness;
5. The physical appearance of basalt and lava were different;
6. The contact relationships of basalt and lava were different;
7. The basalt was divided into regular masses while lava is found as a irregular mass;
8. The dykes of Antrim were different from those seen in Scotland by Hutton;
9. The basalt of Portrush contained fossil marine shells.

The ammonite-bearing basalt was discovered by Richardson in about 1799 and was proof to him and to Kirwan that this lithology was deposited in water. Specimens were sent to Trinity College Dublin, to Kirwan and other geologists, and to Edinburgh in 1801, where they were examined by Sir James Hall, Lord Webb Seymour and John Playfair who recognised their sedimentary nature. The rock is now known to be Lower Jurassic Lias mudstone, which contains specimens of the ammonite *Paltechioceras*. The Edinburgh scientists noted that the Portrush rock had been baked by hot basaltic material (actually dolerite of what is now termed the Portrush Sill) close by, and this was what had made its appearance so similar to that of basalt. On visiting Edinburgh, Richardson was brought to see Arthur's Seat where Playfair tried to persuade him of his error. Richardson held firm in his beliefs.

The Scots' conclusions were confirmed by Jean-François Berger and William Conybeare in 1816, and the scale swung in favour of the Vulcanists, whose view was that basalt was derived from a molten source. Additional evidence came from later observations made in Italy and elsewhere of active and extinct volcanoes where the presence of columnar basalt was seen in unequivocal lava flows, proving beyond doubt that basalt was an igneous rock. Later still the mechanisms of fissure eruptions, which were unknown in the 1790s, were fully comprehended and the origin of the Giant's Causeway more fully understood.

Late in life Richardson turned his pen to agricultural subjects and wrote extensively on the merits of growing fiorin grass. He died in September 1822, continuing to hold to his views on the nature and origin of his 'sedimentary basalt'.

HUTTON RETURNS TO HIS THEORY

Given the initial criticism that his paper of 1785 had provoked, Hutton decided to examine the geology of a number of sites in Scotland. Setting off with John Clerk of Eldin, the engraver and water-colourist, he visited the estate of John Murray, the 4th Duke of Atholl, approximately 75 miles north of Edinburgh, where he wanted to locate the junction between the granite and the mica schist. In the bed of the Tilt, a small river that flows through the steep-sided glen, Hutton found what he was looking for: veins of reddish-coloured granite cross-cutting the mica schist (Figure 6.4). This was conclusive evidence that the granite had been injected in a fluid state from below; it was plutonic.

We should be grateful that Hutton was a friend of John Murray's. Some 62 years later John Hutton Balfour, a cousin of Hutton's who was Professor of Botany in Edinburgh, was in the company of some botany students (who affectionately called him 'Woody Fibre'), and together they tried to visit Glen Tilt in order to examine its plant life. There, on 21 August 1847, they came up against George Murray, 6th Duke of Atholl, and his ghillies, who tried to deny them entry

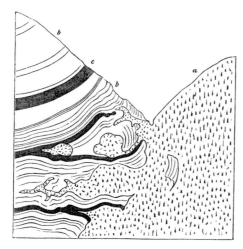

Figure 6.4 Junction of granite and schist at Glen Tilt, Scotland. (a) Granite; (b) Limestone; (c) Schist. Here Hutton found older rocks cross-cut by veins of granite, which proved it was a plutonic rock (from C. Lyell, *Manual of Elementary Geology* (1855), p. 572).

into the glen. Undeterred, they leapt over a dyke and ran down the valley. The Duke tried to assert his right to keep trespassers off his estates, but the 'Battle of Glen Tilt' became widely advertised in song and in print, and the whole affair ended up in court when an Aberdeen solicitor by the name of Abraham Torrie, together with two others, took up the case on behalf of the ramblers. It was declared that a right of way did exist and that 'the pursuers and all others were entitled to the free and uninterrupted use of it'. This action remains one of the most celebrated of Scottish lawsuits.

In 1786 Hutton found similar cross-cutting relationships in granites near Galloway. The following year saw Hutton on the Isle of Arran with John Clerk's son John (1757–1832) (later Lord Eldin) where at Sannex, at Goatfell and in Glenrosa he observed granite veins in the surrounding rocks, and more importantly, at its northern end, he discovered an unconformity. He did not actually call it an unconformity – this term was first used by the Wernerian Robert Jameson in 1805. Hutton later described the feature in his 1795 treatise (volume I, page 429–430):

> I had long looked for the immediate junction of the secondary or
> low country strata with the alpine schistus ...; the first place in
> which I observed it was ... at the mount of Lough Ranza. Here the
> schistus and the sandstone strata both rise inclined at an angle of

about 45°; but these primary and secondary strata were inclined in almost opposite directions ... From this situation of those two different masses of strata, it is evidently impossible that either of them could have been formed originally in that position.

The Arran unconformity, or 'Hutton's unconformity' as it is now popularly known, displays calcareous sandstones that overlie altered and ancient schist (a highly metamorphosed rock type). Hutton did not have to wait long before he found similar examples elsewhere. In the autumn he was walking along the banks of the River Jedd close by Jedburgh in the Borders (40 miles southeast of Edinburgh) and he found another unconformity, this time more obvious. Here nearly vertical beds of a Greywacke (a gritty sedimentary rock) of Silurian age are overlain by horizontal beds of Old Red Sandstone of Upper Devonian age. The actual time between the two is now thought to be least 40 million years. It was this example, rather than that on Arran, that Hutton illustrated in his expanded theory in 1795. In 1788 he was to be found near Cockburnspath near St Abb's Head on the east coast of Scotland, and the weather that day was so poor that he and Sir James Hall were unable to take out the boat they had planned to hire. Instead of viewing a panorama of the coastline they had to content themselves with picking their way along it on foot, and there at Siccar Point they discovered another unconformity (Figure 6.5). As with the Jedburg example Silurian Greywacke is overlain by Old Red Sandstone. Of all three of Hutton's unconformities, the Siccar Point example is the best known, and has, as I said earlier, made its way across the Atlantic as a fibreglass cast. Many of the sites visited were sketched by Hutton's companions, either Hall, or John Clerk of Eldin who executed the majority of the drawings. For a long time these drawings were 'lost' or rather their whereabouts were unknown to the geological community. In 1968 they were relocated at Penicuik, home of Lord Eldin, and in 1978 a portfolio containing many of them was published by the Scottish Academic Press together with an accompanying book authored by Gordon Craig, Donald McIntyre and Charles Waterston.

From a Painting by James Hall, Esq.

Engraved by S. Williams.

Figure 6.5   Unconformity at Siccar Point, Scotland, one of the unconformities recognised by James Hutton. Here vertical beds of Silurian age (a) are overlain by gently aligned, nearly horizontal beds of Devonian Old Red Sandstone (d) (from Lyell, *Manual of Elementary Geology* (1855), frontispiece).

Following this episode of field work, Hutton, instead of committing his geological thoughts immediately into the hands of the typesetters, allowed his interests in chemistry to intervene and he embarked on a number of studies in this area. Two events conspired to get him back to his geological writing. In 1793 he was laid low with a serious medical condition and had to subject himself to the surgeon's scalpel to relieve his tendency to retain urine. As he was recovering from this ordeal he received a copy of Kirwan's paper 'Examination of supposed igneous origin of stony substances' which had just been published in the fifth volume of the *Transactions of the Royal Irish Academy*. It contained a rebuttal of Hutton's 1785 theory. This cannot have pleased him.

That very day, Hutton began composing his response, and it was published in London and Edinburgh in 1795 in book form, in two volumes, as *Theory of the Earth, with Proofs and Illustrations*. The third volume was not published until 1899, when the then Director of the Geological Survey of Great Britain, Archibald Geikie (1835–1924), arranged for its publication by the Geological Society of London. For many years this manuscript had been stored alongside Hutton's two published volumes in the library of the Geological Society of London in Piccadilly where it had been deposited for safe-keeping by Leonard Horner. Perhaps it is this manuscript, or perhaps that on agriculture, or both, that is illustrated piled high on a table cluttered with some fossils and veinose rocks in Sir Henry Raeburn's fine oil painting of Hutton, finished about nine years before the sitter's death.

Hutton's death in 1799 probably resulted in his theory not getting the continued exposure that it deserved. But his reputation was enhanced and his theory advertised widely with the publication of John Playfair's book *Illustrations of the Huttonian Theory of the Earth*, published in 1802 by Hutton's own publisher.

HUTTON COMMEMORATED

James Hutton's house in Edinburgh has now disappeared, having been demolished in the late 1960s, but recently the site has been put to good use. During the bicentenary celebrations of Hutton's death, a plaque was unveiled at the site on Wednesday 6 August 1997, and in 2002 the site was landscaped and opened as a memorial garden, largely thanks to the efforts of Norman Butcher. Incorporated into the garden are several boulders that mark the themes in Hutton's work: conglomerate erratics from near Dunblane illustrate his work on geological processes, while a granite boulder, moved from Glen Tilt, recalls Hutton's pronouncements on granite and plutonism.

While Hutton did not attempt to calculate the date of creation or give a figure for the duration of the Earth, he did open up geology to scientific observation, and removed it from the influence of the church and the reliance for Earth's chronology on the biblical texts. Without doubt his *Theory of the Earth* ranks high, if not highest, in the list of the most influential books ever published in geology. This modest gentleman farmer/geologist deserves the moniker 'The Father of Geology'.

7   The cooling Earth

The twenty-first of September 1792 was a momentous day in Paris. King Louis XVI was deposed and the Republic was inaugurated. Four months later, on 21 January 1793, he lost his head on the Place de Louis Quinze (now the Place de la Concorde) to the newly introduced machine of death, the guillotine – first used in France in April 1792 – and at much the same time the trappings of royalty were dismantled. The royal garden in Paris, the Jardin du Roi, was renamed Jardin des Plantes. This garden was and remains one of the major botanic gardens of the world. It had been laid out in 1626 by Guy de la Brosse and Jean Héroard for Louis XIII, and specialised in medicinal plants – not unusual at the time, when plants were often used in traditional herbal medicine, and when the botanists and medics were always on the look out for new varieties and species which could provide different cures and remedies. The gardens, which now cover 28 hectares, were opened to the general public in 1650, and around them developed a number of institutions which now house the Muséum National d'Histoire Naturelle. Into this museum were brought numbers of exotic species of plants and animals, rocks and minerals collected from the far-flung corners of the French empire that stretched from the east to Louisiana in North America. Following the Revolution the King's own menagerie was removed from Versailles and the animals were transported to the site. In 1870, during the disastrous six-month siege of Paris by the Prussian army, the city folk were forced to eat many of the wild animals. Apparently the elephants were a bit chewy but roast bear was considered good.

The museum quarter in Paris is criss-crossed with streets bearing the names of scientists, zoologists, botanists and anatomists familiar to historians of science. Today the Jardin des Plantes is hemmed in by three such streets (and by the River Seine to the east): to the north by Rue Cuvier, after the anatomist Baron Georges Cuvier (1769–1832); to the west by Rue Geoffroy St Hilaire, after the zoologist Étienne Geoffroy Saint-Hilaire (1772–1844), who in 1827 had accompanied the first giraffe to enter Paris on her journey north from Marseille; and to the south by Rue Buffon after the one-time curator of the Jardin du Roi. It is the latter who is the principal subject of this segment of our examination of the dating of the Earth. Close to the present-day aquatic garden is a bronze statue of the great man, complete with powdered wig and flowing cape, in which he is seated overlooking his garden. In his left hand flaps a dove which looks as if it is desperately trying to escape his grip, and in the plinth it is supposed his heart was entombed.

Near the Jardin, the Université Pierre et Marie Curie is situated on Rue Jussieu, itself named after the three botanical brothers Jussieu: Antoine (1686–1758) who was a Professor of Botany in the Jardin du Roi, Bernard (1699–1777) who forayed throughout the Iberian Peninsula in search of plants and Joseph (1704–1779) who explored in South America and who was responsible for introducing into Europe *Heliotropium*, the heliotrope, an attractive member of the borage family. Later in this narrative the Curies will feature briefly.

THE AVERAGE STUDENT FROM DIJON

Georges-Louis Leclerc (Figure 7.1) was born at Montbard, a small town in Burgundy 200 km southwest of Paris, on 7 September 1707 to Benjamin Leclerc, a state official, and his wife Anne-Cristine Marlin. The family estates lay in the valley of the Armançon River that leads to the Plateau de Langres close to the source of the great French rivers Marne, Meuse, Saône and Seine. Ten years after his birth, Leclerc's father became the lord of Buffon and Montbard, and in that year moved his family to the town of Dijon some 60 km southeast of Montbard. Georges-Louis, who was the eldest of five children, was educated in the town and later began to study law at the behest of his father; such a course of study did not suit this 'average' student and so he moved to

Figure 7.1 Georges-Louis
Leclerc, Comte de Buffon
(1707–1788) (postage stamp
issued by the French Post
Office, c. 1950).

Angers where he began studies in botany, mathematics and medicine.
Leclerc was a bit of a hothead, and like many young members of the
bourgeoisie of his day elected to settle an argument by way of a duel.
Sometimes such clashes resulted in the death of one combatant, and
often one duellist had to leave the district. And this was Leclerc's fate:
following a duel in October 1730, he removed himself as fast as
possible to Dijon. Here he met the young Duke of Kingston, who had
been sent on a Grand Tour by his family in an attempt to teach him
greater maturity. Leclerc joined him on his travels throughout Europe
and returned to Dijon in 1732. His mother died early that year when he
was 25 years old, and when Georges-Louis discovered that his father,
who was fifty, planned to marry a women less than half his age he tried
to wrest his inheritance from him. Leclerc's mother had left her son
her family estates but his father had managed them badly and they
were sold. Fortunately for Leclerc *fils* he was compensated for the loss

and he purchased them back. From then on he lived both in Paris, where he spent the autumn and winter months, and on his estate at Montbard, where he was in residence for the pleasant spring and summer months. At this time he took the additional name 'Buffon' on gaining his estate and styled himself Georges-Louis Leclerc de Buffon. His elevation to that of a Comte, or Count, came in 1773 when the government raised his estates to the status of a 'county'. This change in his personal status came soon after a serious illness that had nearly killed him, and this, no doubt, boosted his morale.

SCIENTIFIC DEVELOPMENT

In Paris he continued his scientific education, and was soon introduced to some of the most brilliant scientific minds in the city. He decided that he should join the Académie Royale des Sciences but to gain admission he had to write some scholarly piece of work. In 1773 he wrote an account of a game he called Franc-Carreau, which was a betting game played with counters on a tiled floor. The contestants bet on the probability that a counter when thrown would land completely inside a tile or would lie across the junction between two adjacent tiles. Buffon discussed the problem of how to ensure the game was fair to all players. He showed that this depended on the proportions of the lengths of the sides of the tiles and indeed their shape. He went on to experiment with changing probabilities caused by throwing different-shaped counters and by 1777 when he finally published this work had added needles to the counter types. This work gave rise to the name 'Buffon's Needle problem' – the first example of study in what is now known by mathematicians as geometric probability theory.

He was admitted to the Académie Royale des Sciences in 1774 as a junior member, essentially an associate, and soon afterwards elected a Fellow of the Royal Society in London. This was surprising because he had not published a great deal up to this time. He engaged in work on the strength of timber, work that was valuable to the French admiralty, and studied chemistry. In July 1739 he was appointed Curator of the Jardin du Roi.

Under Buffon's charge the Jardin du Roi complex was expanded in size and he oversaw a collecting policy that increased the diversity of plants grown on the site and the numbers of plants contained within its herbarium. He built laboratories for scientfic work and glasshouses for the exotic plants and he designed and laid out the maze or 'Labyrinth' as it is still known today. He was not averse to trying to lose visiting dignitaries such as Benjamin Franklin and Thomas Jefferson within its leafy twisting paths. He later clashed with Jefferson over his opinions regarding the nature of the fauna of the New World – in 1766 Buffon had suggested that because of unfavourable climatic conditions, American mammals were less numerous and weaker than their Old World counterparts, and he even went so far as to suggest that Native Americans were weaker than Europeans. Naturally, Jefferson could not agree! In time Buffon became a major player in French scientific circles and was elected to the prestigious Académie Française in 1753, and he remained in charge of the Jardin du Roi until his death in 1788.

HISTOIRE NATURELLE

Buffon is best known for his encyclopaedic *Histoire naturelle, générale et particulière avec la description du cabinet du Roi* that ran to an immense forty-four volumes. The first was published in 1749 and by the time he died, he had published a further thirty-five volumes. Subsequently an additional eight volumes were prepared by Buffon's subdemonstrator in the Jardin du Roi, Bernard Germain Étienne de la Ville, Comte de Lacépède (1756–1825) and others. The series included discourses on zoology, nutrition, classification (he did not accept Carl Linnaeus' (1707–1778) binominal system now in global use), geology and human biology, as well as containing accounts of numerous animal species. It was sensational and ran to at least 52 editions in France and became widely distributed in Europe and North America, with editions in German, Italian, Spanish, Dutch and English appearing in the late 1700s. It proved to be a great money spinner for Buffon who had acquired the financial rights to his work in 1767. Three

English translations appeared between 1775 and 1815: by William Kenrick (1729–1779) and John Murdock (1747–1824); William Smellie (1740–1795) who corresponded with Buffon while he undertook translating the work; and James Smith Barr (*fl.* 1769–1806) who published his translation himself. Jeff Loveland of the University of Cinncinnati has recently remarked that of the three, Barr's was 'particularly careless in pagination and tables of contents. Misspellings were also common ...' Perhaps the most fascinating edition of all was published as recently as 1936. It is illustrated with a series of etchings by Pablo Picasso commissioned by the publishers Vollard, and has been described as one of the artistic masterpieces of the twentieth century.

One reason the series was so successful was that personal paper menageries were now available to those who could afford to own or who could borrow the books. No longer did you have to build up your own collection of wild animals at great expense: you simply lifted them down from their 'cages' on the library shelves and released them when you turned the pages. It must be remembered that the books were published at a time before the advent of public zoological gardens. The word 'zoo' was not coined until 1826 on the establishment of the Royal Zoological Society in London.

Buffon's readable text was wonderfully augmented with engravings of the animals by the artist Jacques E. de Seve, and these set a style for many subsequent and similar zoological compendia. De Seve's animals have an almost statuesque quality: most are pictured on natural plinths of rock in a rather stony stance, while others such as some of the cats are posing on top of ornate pieces of furniture. Whatever we may think of the possible artistic licence taken with the subjects the images do portray the major characteristics of each illustrated species.

A great deal of this work was written at Montbard, where he had established his own private menagerie. These pleasure parks were not uncommon amongst the nobility, but Buffon's was particularly important scientifically, given his work. Buffon restored an old castle and within its desmesne had laid out a garden, built laboratories and stocked a zoological park with many exotic species. There he was able to observe the habits of his animals at first hand, and also carried out some experiments on hybridisation.

BUFFON'S DETERMINATION OF THE AGE OF THE EARTH

Ten years before his death he published his important *Des époques de la nature*, which formed a supplementary part of his *Histoire naturelle*, and it was in this volume that he published his dates for the origin of the Earth. These passages were not his first to deal with geological ideas: in the very first volume of the series published in 1749 Buffon suggested that a comet had crashed into the Sun and this had resulted in some solar material being thrown from its surface. This material separated out into the planets that subsequently revolved around its parental body. In the essay *Théorie de la terre* he outlined his belief that the Earth had a cyclical history that was clearly longer than that given in the biblical accounts. This caused a considerable rumpus and theologians in the powerful Sorbonne demanded that he publish an apology. This he did in subsequent editions, but he did not withdraw his essay.

In the 1760s he returned to geological ideas that were stimulated by his thoughts on thermodynamics or simply heat transfer. He began to explore the idea that the Earth might originally have been molten, and that it was slowly cooling down. This was not a new idea. Isaac Newton (1642–1727) had theorised in his *Philosophiae naturalis principia mathematica* published in 1687 that 'a red hot iron equal to our earth, that is, about 40,000,000 feet in diameter, would scarcely cool ... in above 50,000 years.' Newton suspected that the rate of cooling might vary depending on the diameter of the body losing heat, and hoped that someone might investigate this matter experimentally. Enter Buffon.

If the Earth was cooling down, how long would it take to reach its present temperature? A clue to how Buffon answered this question is contained on a map in an 1819 edition of his *Époques* (Figure 7.2).

Figure 7.2   Map depicting the district around Montbard, showing the location of Buffon's chateau and some iron forges (from Buffon, *Époques* (1819), facing p. 421).

There, marked close to Montbard, are four forges which drew on small deposits of iron ore for their raw material and timber from the adjacent forests for their fuel. One of these was constructed on his estate by Buffon in 1768, and to provide the required power to fuel the workings of the bellows he made his workforce divert a tributary of the Armançon River.

Buffon had begun experimenting with iron in 1767, most probably in a local forge, but obviously enjoyed the experience enough to want his own premises (Figure 7.3). The structure was well built, and amazingly it still stands today (its present owners rent it out to holiday-makers). The best description in English of the structure and ancillary buildings is given in Roger's 1997 biography of Buffon (page 356):

> The blast furnace is remarkable. It is approached by an 'imposing facade,' and a porch surrounded by two alcoves on either side, as if to

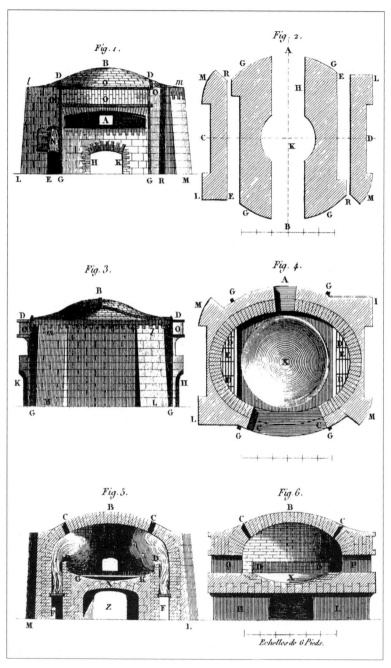

Figure 7.3 Plans of Buffon's forge illustrated in *Oeuvres complètes* IV, p. 118 (from Jacques Roger, *Buffon* (1997), p. 355). Courtesy of Cornell University Press.

shelter statues, which lead to a double-turned staircase with a bannister of wrought iron to the ground level, where the pig iron flowed out. Everything seems to have been built to allow noble visitors to watch the spectacle of 'Vulcan's cave' from above. Behind the blast furnace, two large buildings were devoted to the various activities for the production of iron. Other buildings received the ore brought in carts. Still others contained the lodgings for the permanent workers and the fine dwelling of the steward. Buffon reserved a pavilion for himself where he stayed during his visits.

The last sentence would suggest that he immersed himself totally in his work when he was there.

The forge produced reasonable quantities of iron which Buffon said was the best in the region, and which he sold, but he also used the complex there to carry out experiments on melting and cooling rates of iron, the results of which he interpreted and used to calculate the age of the Earth.

Buffon stated in 1778 what he did:

I caused ten bullets [spheres] to be made of forged and beaten iron; the first, of half-inch diameter; the second, of an inch; and so on progressing to five inches: and all the bullets were made of iron of the same forge, their weights were found nearly proportionable to their volumes.

The spheres were placed in the furnace after which the bellows were driven by two wheels turned by the waters of the Armançon providing oxygen that helped elevate the temperature in the furnace. Once the spheres had been heated to being incandescent (white hot), close to but less than 1,537 degrees centigrade, the melting point of iron, they were placed in a pit and allowed to cool. The temperature in this cooling pit was approximately minus ten degrees which Buffon took to be the actual temperature of the Earth at that time. Alongside the cooling sphere, Buffon placed a sphere of identical material and diameter that had already cooled to the ambient temperature, and this was used to gauge when the hot globe had cooled to the same temperature. This Buffon did by placing one hand on each globe and judging when they were at the same surface temperature. Once he was satisfied they matched each other he read off his watch the time that it had taken the hot globe to reach the temperature of the control globe. This method was not without problems. There was a danger that he could seriously burn one of his palms, and he noted that if the surface texture differed between the two 'bullets' then their respective temperatures were difficult to compare. His results can be seen in Table 7.1.

Being a conscientious scientist he tried to replicate the experiments on the same spheres several times, but failed: with each successive heating event each sphere lost weight as some metal spalled off the surface. Buffon had to rely on the results from the first run in each case.

Extrapolating these results to a globe the size of the Earth, Buffon concluded that 'it would take 42,964 years, 221 days, to cool only to the point where it would cease to burn, and 86,667 years and 132 days, to cool to the actual temperature.' This was considerably more than the 50,000 years that Newton had suggested for the possible duration of global cooling, and completely off the scale of the biblical chronologers!

Did Buffon moderate his timescale in any way? He did, realising that the Earth was not a mass of iron, and that the presence of other materials would affect the cooling rate. In addition, the atmosphere might have acted as a buffer to heat loss, rather like a lagging jacket wrapped around a domestic hot water cylinder, and if the Earth cooled in a vacuum this might have had some effect on the resultant cooling time. Armed with these ideas Buffon carried out heating and cooling experiments on a variety of materials, clay, marls, marble, stone, lead and tin that he had moulded or formed into spheres. His 2-inch sphere of clay took 38 minutes to cool to hand-touch, the one of 2.5-inches 48 minutes, and that of 3 inches cooled in 1 hour 15 minutes. These times are less than those recorded for the iron spheres. He then heated 1-inch spheres of different material close to the melting point of tin

Table 7.1 *Results of Buffon's heating and cooling experiments on metal spheres of various diameters.*

	Heating time to incandescence	Cooling time to hand-hot temperature	Cooling time to ambient temperature
Half-inch sphere	2 minutes	12 minutes	39 minutes
1-inch sphere	5.5 minutes	35.5 minutes	1 hour 33 minutes
1.5-inch sphere	9 minutes	58 minutes	2 hours 25 minutes
2-inch sphere	13 minutes	1 hour 20 minutes	3 hours 16 minutes
2.5-inch sphere	16 minutes	1 hour 42 minutes	4 hours 30 minutes
3-inch sphere	19.5 minutes	2 hours 7 minutes	5 hours 8 minutes
3.5-inch sphere	23.5 minutes	2 hours 30 minutes	5 hours 56 minutes
4-inch sphere	27.5 minutes	3 hours 2 minutes	6 hours 55 minutes
4.5-inch sphere	31 minutes	3 hours 25 minutes	7 hours 46 minutes
5-inch sphere	34 minutes	3 hours 52 minutes	8 hours 42 minutes

(232 °C), and allowed them all to cool down sufficiently so that he could hold them in his hand for half a second. That of iron cooled in thirteen minutes, copper in eleven and a half minutes, Montbard marble in ten, *gres* (a fine grained sedimentary rock) in nine, lead in eight and tin in six and a half. He then worked on other materials including white marble, a soft calcareous stone from Dijon, gold, silver, zinc, antimony, bismuth, gypsum, glass and porcelain, and later deposited the metallic globes in the Royal Collection in Paris.

Analysing all his data he finally concluded that the Earth had taken 2,905 years to consolidate from a molten mass, that following 33,911 years its surface would have been cool enough to touch, and that it reached its present temperature after 74,047 years. His final manipulations gave him a figure of 74,832 years as the age of the Earth. This figure, like his earlier pronouncements in 1749, caused the Sorbonne theologians to complain bitterly and Buffon again published some form of mitigating remarks in later editions of his book.

He then produced a chronological account of the evolution of the Earth which was documented by him in a series of seven *Époques*. The first was the formation of the molten Earth; in the second it had cooled to being just hand-hot; the third was characterised by the world being enveloped in a universal sea; the fourth saw great volcanic activity, followed in the fifth by the emergence of land animals, who somehow managed to survive the tectonic activity of his sixth époque when the configurations of the land masses were defined; in the seventh and final stage, humans appeared. It is interesting to note his organisation of Earth history into seven ages, a number that was congruent with those of the earlier biblical scholars.

After a long and productive life, Georges-Louis Leclerc, Comte de Buffon, died of the painful effects of vesical calculi on 16 April 1788 at home in Montbard at the age of eighty and was succeeded to the estate by his son Georges-Louis-Marie (1764–1794), more widely known as Buffonet. Buffonet was not a successful man despite wealth and connections gained through marriage; he was a spendthrift and not able to settle on any worthwhile study. Catherine the Great of Russia remarked after meeting him that it was ironic that sons of geniuses often turned out to be imbeciles. Buffonet's destiny was to follow King Louis to the Parisian scaffold in 1794.

A few days after Georges-Louis Leclerc's death his body was subjected to a post-mortem examination and his heart removed and given to the geologist and traveller Barthélemi Faujas de Saint-Fond

(1741–1819), a professor at the Jardin du Roi, for burial in Paris. His brain was measured and found to be slightly bigger than normal – he would have been pleased, although perhaps he suspected this fact, given he was nicknamed 'Count Allproud' by some of his contemporaries. The medics also discovered numerous stones or calculi in his bladder which must have been painful and a strain to deal with towards the end of his life. Following a large and lavish funeral in Paris viewed by up to twenty thousand spectators who lined the streets, his body was returned to Montbard where it was laid to rest in the family vault. Later during the troubled times of 'The Terror' that gripped France in 1794 his coffin was stripped of its lead lining, which was melted down for use as bullets: bullets in the sense of ammunition and not in his sense of globes.

Buffon's empirical calculations based on the cooling rate of the Earth rank among the most important measures of the Earth's antiquity, and were taken up with gusto by William Thomson from the 1850s onwards. Of course Buffon did not realise that the Earth has an internal heat source that continues to provide heat and that the temperatures that we now measure are simply not the residual heat left over from its formation. With such a realisation Buffon would have been nearly two centuries ahead of his time. As it was, his determination that the Earth was approximately 75,000 years old was too great a duration for many of his conservative contemporaries to accept.

8 Stratigraphical laws, uniformitarianism and the development of the geological column

By 1800 a new breed of geologist had emerged – the professional. This term includes several groups of people drawn to the discipline: the academics in universities; those who derived their livelihood through working as geologists, mining engineers or surveyors; and those who could support their geological work pretty well full-time through their own means. Into the first of these three categories we can place Adam Sedgwick (1785–1873) of Cambridge (Figure 8.1) and the Reverend William Buckland (1784–1856) of Oxford and later Dean of Westminster; into the second, the army man Joseph Ellison Portlock (1794–1864), and the Local Director of the Geological Survey in Ireland Joseph Beete Jukes (1811–1869); and into the third, Roderick Impey Murchison (1792–1871) (Figure 8.2) and Charles Lyell (1795–1875) (Figure 8.3), to name but six. This emergence generated the momentum that saw the adolescent discipline of geology mature into a fully fledged science, complete with its own professional bodies and surveys and a work force that advanced its understanding and knowledge base. The Geological Society of London was established in 1807 and was followed by other specialist geological societies, including that in Dublin (1831). A chair of Geology was endowed in Cambridge in 1728 by John Woodward, and similar chairs were established in University College London in 1841, and in the University of Dublin in 1843. Government geological surveys began the official mapping of vast tracts of land in the hope of returning economically viable materials. In the developing United States most of the states along the eastern seaboard established surveys: North Carolina in 1824, its

Figure 8.1 Adam Sedgwick
(1785–1873) (from *Geological
Magazine* 7 (1870), facing p. 145).

Figure 8.2 Roderick Impey
Murchison (1792–1871) *c.* 1870.
On the table on his left lies a
copy of his *Silurian System* on
which is resting a geological
hammer. Photograph by Maull
and Polyblank, Piccadilly
(author's collection).

Figure 8.3   Charles Lyell
(1795–1875). Photograph by
John Watkins, 34 Parliament
Street, London (author's
collection).

southern namesake the following year, and New York and also
Pennsylvania in 1836. The first survey responsible for a complete
country rather than a single state was the Geological Survey of
England and Wales begun in 1835 under the dynamic Henry
Thomas De la Beche (1796–1855), its first Director. The Geological
Survey in Canada began work in 1842; in Ireland operations commenced on 31 July 1845; the first state survey in Bavaria, Germany,
began in 1849; the Geological Survey of India began in 1851; Norway
and Sweden 1858; France 1868; the United States 1842; and Egypt
1896. Various museums including the Museum of Economic Geology
in London were also opened around this period. The geologists had
arrived in force.

Given the proliferation of published information from the societies and surveys, the need to place it in some ordered scheme became a

prerequisite to further understanding of the geological history of the Earth. This ordered scheme was stratigraphy, and it provided a coherent logical framework in which the different and recognisable lithological successions could be arranged. The benefits of this were obvious: later researchers could follow on from former workers, and with their newer information, build on pre-existing data, and infill the framework where gaps were present. This all sounds foolproof, but geological societies and their members do not belong to a utopian world, and from the earliest days of these organisations, arguments and downright hostility were commonplace. Individual or closely allied geologists and scientists laid claim to vast swathes of countryside which they felt to be their own domain. Nevertheless, and setting such debate aside, large leaps in the understanding of geological conundrums were made in the 1800s, so that by the 1850s the stratigraphical framework now familiar to modern-day geologists was broadly *in situ*.

STRATIGRAPHICAL GEOLOGY AND MODERN PROTOCOL

Charles Hepworth Holland, who has been closely involved with matters of stratigraphy, particularly that relating to the Silurian, defined stratigraphy thus in his book *The Idea of Time* (Chichester: Wiley, 1999): 'Stratigraphy is the study of successions of rocks and the interpretation of these as sequences of events in the history of the Earth.' Stratigraphy is one of the fundamental disciplines of geology and has been the focus of huge volumes of research. Today many matters of stratigraphical protocol are controlled by the International Union of Geological Sciences (IUGS), which has a number of committees or commissions to oversee particular geological periods/systems or boundaries between them. For example we have, or had until their work was completed, the sub-commissions on Devonian Stratigraphy, and the Silurian–Devonian Boundary, to name but two. These commissions are formed of geologists from around the world who have a particular interest in rocks of a particular age. While geologists are fundamentally interested in rocks as a whole, many academic and professional geologists focus on one or perhaps two parts of the geological column – to develop an encyclopaedic knowledge of the complete span would be virtually impossible. Today the discipline of stratigraphical study is split into three sub-branches: lithostratigraphy, biostratigraphy and chronostratigraphy. Lithostratigraphy is concerned with the naming, defining and description of rock units in terms of their physical characteristics. Distinctive units that can be marked on a map are termed *formations*, and several in close proximity to each other may be bundled together into a *group*. In biostratigraphy fossils are used for correlation between districts, and *biozones* are the formal units that are characterised by a particular fossil content. Chronostratigraphy concerns itself with the definition of internationally agreed boundaries between units known as *systems*, *series* and *stages*, and these boundaries are marked at particular points in appropriate geological horizons (stratotypes) and are each known as a 'golden spike' whose absolute geological age is known. A 'System' comprises all the rocks between its lower and upper boundary and it corresponds to the division of geological time known as a 'Period'. Thus rocks of the Silurian System were deposited during the Silurian Period.

Such formality in dealing with stratigraphy has not always been the case, and here we are primarily concerned with the establishment of the divisions of geological time at a time when absolute ages were not known. As has been discussed earlier, many geologists in the eighteenth century followed the three-fold division of the rocks into Primary, Secondary, Tertiary, while some included Quaternary as a fourth division, a classification that owed much to the effects of the Great Flood associated with Noah and the Ark. However, by the early 1800s the biblical interpretation of the formation of the geological succession was largely discarded, except by some die-hard believers in the Mosaic story. Dominick McCausland (1806–1873), the Irish barrister and author of books such as *The Times of the Gentiles*, published a geological column as a frontispiece in his popular *Sermons in Stones*. The geological periods are placed not in Eras but in Days, with the Cambrian within the first day and the Tertiary in the sixth and final day. Even today, there are communities who still cling to such beliefs.

From about the first decade of the eighteenth century, geologists and gentlemen of science began to examine the sequences of rocks in far greater detail than had their predecessors, and soon they realised that the earlier classification was too simple, and that succession was far more complex. Lithologically the rock types were quite varied, and their origins could not be ascribed to earlier interpretations. Within Europe a number of schemes that gave descriptive terms to various distinctive portions of the geological succession were promoted, and there was no clarity or accepted terminology. This meant in practice that it was difficult for a French reader of a paper published in England to appreciate fully which horizon was being written about, and to correlate it with continental horizons with which he might have been familiar.

In Britain the first descriptive table of strata was published as early as 1719 by John Strachey, the mineral prospector and surveyor. Strachey tabulated and named the succession in ascending order of age: Mineral-bearing rocks below the Coal Measures, Coal Measures, New Red Sandstone, Lias, Oolite, Chalk. He later expanded this work and produced a pamphlet entitled *Observations on the Different Strata of Earths and Minerals, More Particularly of Such as are Found in the Coal-mines of Great Britain*. At many large geological congresses held today a glossy card printed with the geological column will be available; often the size of a credit card, it fits easily into a wallet, and can be surreptitiously whipped out during a Friday night table quiz down at the local pub, or equally easily by a forgetful geologist fearful of being embarrassed in front of his or her peers. If you compare Strachey's efforts with this geological column, immediately it will be clear that some terminology is common to both, and that the arrangement of Strachey is somewhat familiar. In essence he compiled an early version of the modern-day geological column. Similar tabulations were erected by others, and terminology proliferated.

John Michell, in his paper on earthquakes, also discussed the nature and extent of various geological units in England. These he tabulated and named in a manuscript dating from 1788; the list was published posthumously by the mineral surveyor John Farey in 1810. However, it contained stratigraphical nomenclature that is largely unrecognised today.

Clarification and general acceptance of stratigraphical units were to come later, by the 1840s. Certainly within Britain, as field work gathered pace and results were disseminated, geologists began to recognise the clear differences between the red sandstones that occur in north Devon, which were informally called the Old Red Sandstone, and the finer-grained silts, slates and greywackes that occur on the other side of the Bristol Channel in south Wales. They also began to recognise the relationship between varying lithologies and realised that, for example, the Coal Measures and coarse grits, called the Millstone Grit on account of their use as quernstones and millstones, that occur in Lancashire and Yorkshire directly overlie the grey crystalline limestones familarly seen in the Pennines. They recognised that the green chloritic sands seen near Cambridge, naturally termed the Greensand, are older than the chalk so splendidly and dramatically exposed along the Dover coastline.

The vertical spatial relationships on a local level were quite easily unravelled, but geologists went further and began to determine the relationships between rock successions much further afield. The sandstones of Devon were recognised to overlie the silty greywackes of south central Wales, but were overlain themselves by the crystalline limestones found in Yorkshire and elsewhere. Through detailed observation and description of British geology, a number of geological Periods were proposed in the early 1800s, all of which represented a portion of geological time, and all based upon distinct lithological and palaeontological grounds. As similar work was carried out in Europe and in Russia, other distinctive lithological units were described and ascribed to newly named geological Periods – terminology that began to replace the older more descriptive terms such as Old Red Sandstone,

Oolite, and Lias. In this way the geological time units – the Eons, Eras, Periods and Epochs – were discerned, and the geological column as we know it today was born.

LAYING DOWN STRATIGRAPHICAL LAWS

In the period between 1800 and 1841 geology became formalised as a science. While in the past observers often made random pronouncements on what they had seen, there were few underlying principles to guide their thought processes. It was the work of two men, William Smith (1769–1839) (Figure 8.4) and Charles Lyell, that largely laid down the laws and principles of stratigraphy and correlation on which understanding and full appreciation of geological processes and their effects depend. At the same time, four major strands that reflected this growing professionalisation began to emerge: firstly, there evolved an appreciation that time could be considered in geological terms; secondly, the stratigraphical framework in which all modern geologists work was erected on a firm foundation; thirdly, geological societies were established as centres of national and provincial learning; and finally, systematic mapping of geology of various countries was placed on a sound footing and national geological surveys were constituted to carry out such field work.

WILLIAM SMITH

William Smith, who was the uncle of the geologist John Phillips (1800–1874), is now known as the 'Father of English Geology'. Born in Churchill in Oxfordshire, he had a varied career as a canal engineer, as a mineral surveyor and later as a land steward. He became involved in the surveying and construction of canals in Somerset, at a time when this method of transport was gathering pace in the early years of the Industrial Revolution. As canals were cut across the English countryside, Smith became very acquainted with the geology. He soon began to map the distribution of various horizons, and in 1799 produced a hand-coloured manuscript geological map of the area around Bath and a manuscript table of strata. Two years later he produced a

Figure 8.4 William Smith (1769–1839) (from Horace B. Woodward, *The History of the Geological Society of London* (London: Longman Green & Co., 1908) facing p. 92). Courtesy of the Geological Society of London.

rough geological map of England and Wales. Eventually in 1815, he produced his masterful geological map entitled *A Delineation of the Strata of England and Wales, with Parts of Scotland*, which was printed at the large scale of five miles to an inch. This work was one of the most important geological publications ever produced. In 1819 he published various cross-sections graphically illustrating the underlying and thus three-dimensional geology of his 1815 map. Between 1819 and 1824 Smith also produced a series of geological maps of twenty-one English counties including Cumberland, Gloucestershire, Oxfordshire and Wiltshire, contained in seven folios that built up into a comprehensive atlas. These were all published by John Cary of London.

Smith was not always appreciated by his peers, and he endured great hardships during his lifetime. He was never a member of the prestigious Geological Society that had been founded as a dinner club in London in 1807. Perhaps this was because he was only a mineral and canal surveyor, a breed of professional, and as a result was looked down upon by many in the gentlemanly classes that dominated the

English geological establishment. Perhaps also Smith's period of imprisonment for debt did not help his advancement. Nor did it help that in 1819 the Geological Society published a large-scale geological map of England and Wales which was remarkably similar to his 1815 map. This 1819 map was prepared by George Bellas Greenough (1778–1855), one of the society's leading lights and its first President. Towards the end of his life Smith received two geological accolades, one of which, the award of a Wollaston Medal by the Geological Society, must have seemed somewhat ironic to him. The second honour must have given Smith some pleasure: the conferring of an honorary degree in law, not by an English university, but by Trinity College Dublin, during the 1835 meeting of the British Association for the Advancement of Science, which was held outside mainland Britain for the first time that year.

Smith's two ground-breaking geological laws were the Principle of Superposition, and the Law of Strata identified by fossils, both of which were formulated in his publication *Strata Identified by Fossils*, published in four parts between 1816 and 1819. The first Law said that in a sequence of beds of rock, those that lie on top are younger than those below, unless there is clear evidence to suggest that the whole succession has been overturned. Such reversal could be caused by folding and/or faulting of the rocks. Smith's second Law said that each bed contained a distinctive fossil assemblage. These ideas were important as they allowed geologists to appreciate the original geometry of geological successions, and also to be in a position to correlate horizons for long lateral distances even if there was a break in the outcrop of that horizon on the surface. In terms of drawing geological maps, these laws were fundamental, and an understanding of them made the task simpler.

CHARLES LYELL

Charles Lyell came of quite different stock from Smith. Born in the family seat Kinnordy House, close to Forfar in Scotland, the centre of an estate that is still in the Lyell family, the young Charles divided his childhood between the family home in Scotland, and another leased by this father near the New Forest in Hampshire in England. He studied classics and mathematics at Exeter College, Oxford, but later turned his attention to geology, after his reading in 1816 of Robert Bakewell's (1768–1843) *Introduction to Geology*, which he had found on the shelves of his father's library. Such was the influence of this book that Lyell took time out from his premier studies to attend the lectures of the Professor of Geology, the flamboyant William Buckland, who later was to engage in a close study of coprolites (fossil faeces) produced by the marine reptiles found fossilised in the Jurassic successions of southern England. Immediately, and unsurprisingly given Buckland's dynamic enthusiasm, Lyell was gripped by this science that was new to him. Evidence of this newly found devotion is documented in a series of letters written by him while still an undergraduate at Oxford, in which he jotted down observations on the geological nature of areas through which he had travelled on his return to Scotland. After Oxford he embarked on a tour of France, Switzerland and Italy in the company of his family, during which he honed his geological ideas. Lyell was fluent in several European languages, which allowed him to interact and correspond most effectively with the leading continental geologists of his day. In this respect he was not isolated geologically from their influence as were many of his British contemporaries.

After a brief period as an academic in London (he had been appointed as Professor of Geology at King's College, London in 1831 but resigned his chair two years later) Lyell occupied his time as a gentleman of science and travelled extensively in search of geological evidence. He was closely associated with the Geological Society in London where he served in several posts including, twice, that of President. Lyell was somewhat of a geological maverick; he popularised the subject through his highly successful textbooks and delivered lectures to huge audiences on both sides of the Atlantic. His major claim to fame lay in the success of his two great textbooks *Principles of Geology*, which first appeared in 1830 and which went through many editions, and *Elements of Geology* that appeared eight years later. *Principles* was an immediate bestseller for John Murray its publisher, who sold 15,000 copies to a receptive audience. It must be remembered that geology in the 1830s was emerging as an exciting new science at a time when travel had become possible to a sizeable portion of the population. Gentlemen as well as others were keen to embrace geology, many became involved with local and provincial scientific and literary societies and were anxious to gain a rapid understanding of the science. To them, the acquisition of Lyell's works was therefore of great importance, and Lyell exploited his popularity by undertaking four lecture tours in North America between 1841 and 1853. These trans-Atlantic sojourns were rewarding in two ways. In Boston in late 1841 he delivered twelve lectures to an average audience of 3,000 and received $2,000, equivalent to $30,000 today. He also recorded the geology of many parts of the eastern North American continent, observations that he subsequently worked into later editions of his books.

As we know, a major stumbling block to geological thinkers was the Bible, and the limited amount of time that it made available to produce geological phenomena. Geological events were viewed as being catastrophic in origin: one-off episodes that were responsible, for example, for fossils, sedimentation and erosion. By dispensing with this catastrophism constraint, geologists were able to broaden their imagination and vision. Suddenly, the volcanic rocks described by John Strange in northeast Naples and by William Richardson in northeast Ireland could be seen to be similar in origin to those produced by contemporary volcanoes such as Vesuvius and Mount Etna. Basalt was not the product of sedimentation from the Noacean Deluge, but was volcanic in origin, and more importantly could have been erupted at any time in the past. As was propounded forcefully by Lyell, the past history of the Earth could be explained by contemporary Earth processes: or as he put it 'the present is the key to the past'. This statement became enshrined as the Law of Uniformitarianism, and remains fundamental to the understanding of the dynamics of Earth history. Lyell also realised that geological events could take place reasonably quickly but smoothly – a realisation that found its most forceful illustration at the 'Temple of Serapis', a ruined Roman commercial edifice situated at Pozzuoli close to Naples, an engraving of which Lyell chose to use as a frontispiece in many editions of his *Principles* (Figure 8.5). At Pozzuoli three columns of the entabulature remain. Curious holes decorate the columns up to a height of twenty-one feet above the present high water mark. Lyell visited the site in early 1830 and deduced that the columns had become partially submerged and the borings had been produced by the extant marine bivalve *Lithodomus* in the recent past before the columns re-emerged from the ocean. This marine transgressive–regressive sequence could have been due to a lowering and rising of the land surface, or of the sea level, or both. Whatever the cause, this dynamic movement did not cause the remaining columns to fall and so was relatively smooth and fast. This uniformitarianist view followed Hutton's and was in contrast to the views of some geologists such as Adam Sedgwick who held that many geological events were sudden and catastrophic.

Lyell was hugely successful as a geologist, and was widely known, both in Europe and in North America. On his death he was accorded the honour of being found a burial place in Westminster Abbey. While Lyell was perhaps the public face of geology in Britain at this time, his ideas were not universally accepted, particularly amongst some frequenting the tight circles of academia and the Geological Society. Sedgwick did not subscribe to all his views, nor did Buckland.

THE CONCEPT OF GEOLOGICAL TIME

As geologists recognised distinctive lithological units or groups of units which could be distinguished from each other through examination of their fossil content, those same geologists began to realise that each distinctive unit must represent some time in the past. Geological time, as an abstract concept, began to unfold in the early decades of the 1800s, and while absolute timescales could not be attributed to the emerging geological periods, the understanding that actual time was

Figure 8.5   The 'Temple of Serapis' at Pozzuoli close to Naples (from C. Lyell, *Principles of Geology*, 7th edition (1847), frontispiece).

represented in the geologial succession was most important. In a recent volume published by the Geological Society of London to celebrate the bicentenary of the birth of Charles Lyell, Joe Burchfield discussed the conceptual development of geological time. It required, he noted, five essential ingredients or steps: the geological succession had to show evidence of past events; scientists had to accept that the Earth was older than the historical record; a sense of historical antiquity developed through the erection of the geological column; methods had to be designed to quantify the actual duration of geological time; and finally scientists had to accept that there was a finite limit of time. With all of these there came the sense that geological time was perhaps far longer than hitherto appreciated, that this 'deep time' could be deciphered through an examination of the rocks, and that through this a chronology of terrestrial history could be presented and comprehended.

GENESIS OF THE GEOLOGICAL COLUMN

Without wishing to sound Eurocentric, it can be said that stratigraphic geology was founded in the tiny geographical area of Britain and parts of Europe. This region has yielded a largely uninterrupted sequence through the fossiliferous rocks. However, standard geological nomenclature has proved difficult to apply in other parts of the globe: for example, not all examples of red sandstones found in the United States, Australia or Asia, can actually be correlated with the red clastic rocks of Devon, and they may actually occur in different geological Periods.

Today the geological column (see Frontispiece) is divided into four major divisions or Eons. The earliest is called the Hadean, named by the American stratigrapher Preston Ercelle Cloud Jr (1912–1991) after Hades, the hell or underworld of Greek mythology, on account of the fiery condition of the Earth from its initial formation to the formation of crustal rocks and the accretion of proto-continents. Little information on this Eon is known from Earth, but the Moon has provided a great deal of data on this the earliest 700 million years of our history. Next come the Archean (from the Greek for primitive), the

Proterozoic (first life) and the Phanerozoic. The Archean, which was named by the American mineralogist James Dwight Dana (1813–1895) in 1872, and the Proterozoic, named by another American geologist Samuel Franklin Emmons (1841–1911) sixteen years later, comprise the immense pile of largely unfossiliferous rocks that were produced or deposited between the development of crustal rocks formed by the differentiation of the Earth's material into a central core, middle mantle and surface crust, and the point at which marine animals were able to precipitate hard shells. The youngest Eon is the Phanerozoic, a term coined from the Greek *phanero* and *zo*, meaning visible and life respectively, where lithological variation is greater than before and where life on Earth began to diversify, and at times wax and wane.

The Phanerozoic in turn is composed of three Eras – Palaeozoic (ancient life), Mesozoic (middle life) and Cenozoic (recent life). The first was named by Adam Sedgwick in 1838 in a paper published in the *Quarterly Journal of the Geological Society* (of London) to encompass the two lowermost geological Periods. However, just two years later John Phillips redefined it to include the younger Devonian Period. It now comprises six geological Periods. The Mesozoic and Kainozoic (now more often written Cenozoic) were so named by Phillips in 1840 in a paper published in the *Penny Cyclopedia*, a widely distributed popular magazine that did much to spread the geological word in Britain. The Eras are themselves divided into Periods, which are shorter time-spans often characterised by distinctive rock types and recognised by their particular fossil constituents. It is with this level of subdivision of the geological column that people are most familiar, having heard on television of the 'Cambrian explosion', or seen the blockbuster movie *Jurassic Park*. Nearly all Periods are themselves subdivided into lesser Epochs, Stages and Series, and even into Zones, which can comprise a very short span of time, and are recognised by just one diagnostic fossil.

By and large the Stages and Series are named in a parochial way, often reflecting an area where the rocks occur, and it can be difficult when reading the geological literature to correlate packages of geology that carry different monikers over long distances. In the United States a whole suite of names has been applied to Ordovician stages that many European geologists would find as difficult to decipher as an American would Serbo-Croat: Canadian, Chazyan, Blackriverian, Trentonian and Cincinnatian. These are approximately equivalent to the Tremadoc, Arenig, Llanvirn, Llandeilo, Caradoc and Ashgill stages in Britain. The answer to stratigraphical definition is to apply global standards and to state precisely the limits and equivalents of geological time packages however small or large. Approximately twenty years ago global standards were published for the Silurian Period but even these standards are now being questioned by younger geologists. The geological column continues to evolve.

WHITHER THE GEOLOGICAL PERIODS?

Who defined and introduced the names of the various geological Periods? In order to appreciate the history and timing of these pronouncements it is worthwhile answering this question in chronological order. In all there are now twelve geological periods, although there has been debate among some stratigraphers on the status of the subdivisions of the Tertiary, which itself has recently been consigned to the geological waste bin along with the Quaternary. They have been replaced with the Paleogene and Neogene, much to the regret of historians of geology who remembered the connections between stratigraphical terminology and the pioneers such as Arduino. These two names were first used as terms for two sub-Eras contained within the Tertiary, and their authors, M. Hornes who in 1853 coined the term 'Neogen' [= Neogene] and C. F. Naumann who in 1866 first used the term 'Paläogen' [= Paleogene], never imagined that they would attain Period status.

To non-geologists, the geological column and the order of its Periods is somewhat difficult to remember, assuming that it is information that they would want to remember in the first place! Most geologists would be able to rattle off the names of the geological Periods in the correct order with ease. Various mnemonics have

Table 8.1 *The authors of various divisions of the geological column, and the dates the terminology was first used.*

Era	Period	Epoch	Date	Author
Cenozoic			1840	John Phillips
	[Quaternary		1829	Jules Desnoyers]
	Neogene		1853	M. Hörnes
		Holocene	1867	Paul Gervais
		Pleistocene	1839	Charles Lyell
	[Tertiary		1760	Giovanni Arduino]
		Pliocene	1833	Charles Lyell
		Miocene	1833	Charles Lyell
	Paleogene		1866	C. F. Naumann
		Oligocene	1854	Heinrich von Beyrich
		Eocene	1833	Charles Lyell
		Paleocene	1874	Philipp Wilhelm Schimper
Mesozoic			1840	John Phillips
	Cretaceous		1822	Omalius d'Halloy
	Jurassic		1795	Alexander von Humboldt
			1839	Leopold von Buch
	Triassic		1834	Friedrich August von Alberti
Palaeozoic			1838	Adam Sedgwick
	Permian		1841	Roderick Murchison
	Carboniferous		1822	William Conybeare and William Phillips
		Pennsylvanian	1891	Henry Shaler Williams
		Mississippian	1869	Alexander Winchell
	Devonian		1839	Adam Sedgwick and Roderick Murchison

Table 8.1 *(cont.)*

Era	Period	Epoch	Date	Author
	Silurian		1835	Roderick Murchison
	Ordovician		1879	Charles Lapworth
	Cambrian		1835	Adam Sedgwick; named by Murchison
Precambrian			1862	Joseph Beete Jukes

been penned by unknown authors which help one to remember the sequence, but given the recent modifications to the geological column they do not work. The author remembers being taught the Epochs of the Tertiary and Quaternary by bringing to mind the fantastically named imaginary Tertiary pachyderm EOMPPR: Eocene, Oligocene, Miocene, Pliocene, Pleistocene and Recent. The creature has now been reduced to cerebral extinction.

Although most divisions were named and defined in the 1830s, a number pre-date this burst of nomenclatural activity in the geological literature (see Table 8.1). The earliest, the *Tertiary*, is a throwback to the work of Giovanni Arduino, who applied the term in his four-fold division of rocks in Italy, and while he did not rigidly define the limits of the Tertiary, it became used for rocks deposited after the Chalk and before the deposition of the loose frosting of the alluvial drift. The *Jurassic*, which is perhaps the geological Period most familiar to the general public through those Hollywood dinosaur romps, was first used in 1795 by the celebrated German scientist Alexander von Humboldt in his description of successions exposed in the Jura Mountains of Switzerland. Humboldt was a scientist with an all-encompassing interest in many branches of learning, and although he is known for his monumental description of the physical characteristics of the world, published as *Kosmos* or *Cosmos*, in seven or more volumes, and in several translations between 1845 and 1867, and remembered today in the name of the ocean current that laps against the Andes in South America, his geological research is not generally well known. Humboldt's terminology was later more rigidly applied to a more precise region and narrower timeframe in 1839 by Leopold von Buch, a German geologist and mineralogist, who also wrote a fine travelogue on Norway and Lapland in 1810.

Chalk is perhaps the lithology or rock type most instantly recognised by most people. We are aware of it from an early age through its use on squeaky blackboards in dusty, noisy classrooms. Features on the English landscape that are composed of chalk are all familiar: the creamy white Cliffs of Dover stand as a step into England from the Continent or as a metaphorical defence from it; the White Horse, and the Giant with his club held above his head are cut into the underlying white rock. The lighthouse at Beachy Head is perched immediately on top of chalk cliffs, which being composed of this soft pure limestone unfortunately erode rather too rapidly for those who own property near by. The chalk was deposited during the Cretaceous, a period defined in 1822 by the Belgian geologist Jean Baptiste Julien d'Omalius d'Halloy (1783–1875). Unlike many contemporaries who carried out geological field work on horseback, d'Halloy preferred to walk everywhere. He is principally known for his work on the younger rocks and fossils exposed in the Paris Basin, and for his geological map of much of western Europe, the culmination of his decade of geological perambulations. He was also an educator and penned a number of early geological textbooks including *Abrégé de géologie* which proved highly popular and ran to at least seven editions.

The Reverend William Daniel Conybeare (1787–1857) was an English clergyman, whose early ecclesiastical career saw him ministering to parishioners in Bristol, before he headed across the River Severn to Sully in south Wales in 1824. His Welsh sojourn lasted nine years before he was moved to Axminster, now noted for its carpets. He reached the pinnacle of his clerical career in 1845 with his elevation as Dean of Llandaff in Cardiff, where he remained until his death. If you wander around Llandaff Cathedral and make your way across the neatly kept lawns to the nearby Chapter House, you may find his final resting place. It bears little witness to Conybeare's earlier activities as a geological and palaeontological pioneer.

Like many men of learning at this time Conybeare was heavily involved with the local scientific and literary society, the Bristol Institution, which opened its doors in 1809. He became interested in ichthyosaurs and plesiosaurs, a group of wonderful marine reptiles found in the Liassic rocks in Somerset and on the southern coast of England at Lyme Regis where they were collected and brought to the awed attention of the general and scientific communities by Mary Anning (1799–1849). What were these animals like? Superficially ichthyosaurs resembled modern dolphins: they were streamlined, had four paddles, two front and two rear, and had a powerful posterior fin. The head carried a huge eye on either side, while the snout, which was drawn-out and elongate in shape, carried upwards of two hundred simple peg-like teeth. Plesiosaurs were broadly similar except that they possessed a very long neck, which in some species was nearly half the length of the body. These physical attributes would have made these animals fierce predators in the Liassic oceans, and they are known to have eaten fish and ammonites. A close examination of coprolites, the fossilised remains of their dung, provides conclusive proof of this. Although long gone, representations of these animals stand in the grounds of the Crystal Palace in Sydenham in south London, where people can see for themselves and marvel at the remarkable reconstructions of various prehistoric animals. Perhaps the most famous of these is the model of *Iguanodon* in which twenty-two eminent Victorian scientists and palaeontologists enjoyed a sumptuous dinner on New Year's Eve 1853, hosted by Benjamin Waterhouse Hawkins (1807–1899), the sculptor who had fabricated the models.

Ichthyosaurs and plesiosaurs had been first found in Britain as early as 1605, but it was not until the celebrated Lyme fossil collectors Mary and her brother Joseph began to find complete skeletons in the local cliffs, from about 1811 onwards, that the scientific community

sat up and really took notice. Although attempts had been made earlier to describe the physiology and osteology of these streamlined
beasts, it was 1821 before clear descriptions of their structure were
published by Conybeare and Henry De la Beche in the *Transactions of
the Geological Society*. Conybeare followed this paper up with a
second published the following year in which he described and differ-
entiated four species of ichthyosaur based on differences in the morphology of the teeth.

However, in the context of the present narrative, 1822 is also a
significant date, as he and William Phillips (1773–1828) published
their *Geology of England and Wales* as an accompaniment to their
geological map published a year earlier. This volume attempted to
place the geology they described in the context of that seen elsewhere
on the Continent. They were well placed to produce this work, as both
were geologists of considerable ability. Phillips was a founder member
of the Geological Society and a Fellow of the Royal Society and had
already published a volume and contributed to another, on the geology
of England and Wales. He had another valuable asset: he was a printer
and bookseller, so the cleric and typesetter had no difficulty finding a
publisher and distributor for their work. In their book Conybeare and
Phillips erected a hierarchy of orders: the granite was placed in the
Inferior, the bulk of the stratified successions were assigned to one of
three medial orders while the overlying unconsolidated sediments
were placed in their Superior order. Within the Medial order they
defined the Carboniferous: this term had been first suggested by
Richard Kirwan. In the type area in Britain three lithologies were
recognised: the Mountain Limestone was the oldest and comprised
flat highly fossiliferous bedded limestones; overlying this were the
Millstone Grits, a succession of coarse sandstones; and these were
topped off with the Coal Measures, a sequence of coal seams, sandstones and shales which were primary commodities that fuelled the
Industrial Revolution in Britain.

The term Carboniferous is now used globally. Until recently geologists in North America recognised two Periods, the Mississippian

and Pennsylvanian, that equated to the Carboniferous, but recently agreement has been reached by the members of the Sub-Commission of Carboniferous Stratigraphy of the International Union of Geological Societies, who have agreed that they become sub-series of the Carboniferous. Application of this edict globally should allow for greater clarity.

The Mississippian, which is the lower half, and which correlates more or less with the Carboniferous limestone, was defined and introduced in 1869 by Alexander Winchell (1824–1891). Winchell was a polymath who was appointed Professor of Physics and Civil Engineering at the University of Michigan in 1854, but switched subjects a year later to become Professor of Geology, Zoology and Botany. He also served as the State Geologist of Michigan during two periods, 1859–1861 and 1869–1871, and was joined by his brother Newton Horace in 1869 as his assistant. For his state geological surveys, which were interrupted by the Civil War, Alexander received a salary of $1,000 for six months' work, but his tenure was to end in turmoil with the refusal of the authorities to publish the report of his second survey, and his subsequent resignation. He later moved to Vanderbilt University where his tenure ended when his lecturing position was abolished. Winchell had been unwise enough to publish an article that the authorities decided promoted the idea of evolution: for this he had to go. He returned to the safer confines of Michigan.

The upper half of the Carboniferous in North America, which is broadly equivalent to the Millstone Grit and the Coal Measures, is known as the Pennsylvanian, a tag first used in 1891 by Henry Shaler Williams (1847–1918) for the coal-rich successions of Pennsylvania. Williams was born in New York State and received his geological education at Yale. He subsequently became Professor of Palaeontology at Cornell University. Not only did he publish on stratigraphy and palaeontology, but he also published a celebrated treatise entitled *Bones, Ligaments, and Muscles of the Domestic Cat*. One can imagine the panic in the feline fauna as Williams walked the alleyways of Ithaca. At Cornell he became involved as third President from 1895 until 1901 with the student honour society Sigma Xi, which was established in 1886. Although similar honour organisations existed for other subjects and in other universities, Sigma Xi was the first of its kind in science. To many non-Americans these organisations with their Greek names sound suspiciously like some religious cult. This is not the case. Sigma Xi was set up as a scientific honour society to encourage members to develop a sense of belonging and cooperation in scientific and engineering research, and to reward subsequent excellence. Rapidly the organisation expanded so that by 1900 it boasted over a thousand members in eight chapters. It has to be remembered that many universities in North America were rather young compared with those in Europe, and so the proliferation of honour systems helped to promote a sense of loyalty amongst alumni. After his retirement from Cornell, Williams spent a great deal of time in Cuba working on various oil prospects. He died in Havana.

The two decades of the 1830s and 1840s saw the defining of most of the remaining geological succession into clear and recognisable Periods or lesser units. In 1833 Charles Lyell published a subdivision of the Tertiary into several Epochs based on the number of living species that they contained. His oldest Epoch, the Eocene, from the Greek *éos* meaning dawn, contained a small number of extant species; the Miocene, from *meîon* meaning less, contained rather more; and his youngest unit, the Pliocene, from *pleîon* meaning more, had more still. All the sediments younger than Pliocene, which were coeval with human activity, he placed into the Recent, although six years later he recognised that the lowermost sediments contained some extinct animals and he termed this older unit the Pleisocene. Today it is most associated with the last ice age, a geological phenomenon not recognised in the 1830, and its characteristic fauna of woolly rhinoceros, woolly mammoth and giant Irish deer. It is hard to believe that as recently as 20,000 years ago these animals roamed the area around modern-day Trafalgar Square. Did they celebrate the passing of the old year with such enthusiasm as do modern revellers?

In Wurtemberg, the German geologist Friedrich August von Alberti (1795–1878) was interested in the successions exposed around the town, and spent considerable time looking for minerals of economic value. In doing so he recognised in 1834 that they could be subdivided into three units, the Buntsandstein (known as the Bunter in English), the Muschelkalk and the uppermost Keuper, collectively known as the Trias. The Muschelkalk was so named on account of its containing a large number of fossil shells. Modifying long-established stratigraphical nomenclature is not popular amongst geologists. Today rocks of the Triassic are extremely well known as they are found in the oil fields of the North Sea and in western Europe, where the porous horizons act as reservoirs for either oil or gas, and the marls horizons as seals on oil wells which stop the valuable commodities seeping and escaping into overlying sediments. The Triassic in onshore Britain is dominated by terrestrial sandstones of the New Red Sandstone, but it also contains considerable reserves of salt and other evaporite minerals, such as gypsum, which are extracted for use in the building trade and in the chemical industry. Next winter as you drive along treacherous icy motorways you will no doubt curse the local authority for being slow to spread salt – Triassic salt – on the road surface.

In 1830 two explorers began busily mapping vast tracts of land – no, they were not working their way through parts of the mid-West of America, nor walking through the virgin outback in Australia, but were in Wales, a landscape that had been highly modified by the action of man for at least seven millennia. These geological explorers and great friends, Adam Sedgwick and the highly ambitious Roderick Impey Murchison, were engaged in mapping the northern and central parts, respectively, of the principality.

Sedgwick was a Dalesman, born and bred in the village of Dent in Cumbria where a large monolith of beautiful Shap Granite now stands in his memory and slows down traffic negotiating the narrow village street. Close by this speed deterrent is the Old Rectory where Adam was born on 22 March 1785. Murchison was a soldier, and a veteran of the Peninsular Campaigns, before becoming a late convert to geology, thanks to his wife's influence, following his discharge from the army. Like many army engineers, he was well positioned to take up the geological challenge Wales offered. He was organised and meticulous, and embarked on his work with the zeal of a general attempting to get to Moscow before winter.

It was clear that the deformed slaty rocks of northern Wales, which are so beautifully exposed in the Snowdonia region, were older than the highly fossiliferous siltstones and overlying limestones found in the Welsh Borderlands. Sedgwick named his tract 'Cambrian', on the suggestion of Murchison, who gave his lower succession the name 'Silurian', after the tribe the Silures. However, as they began to publish the fruits of their field work in papers and in books both Sedgwick and Murchison began to expand the frontiers of their territories so that they began to overlap. The precise attribution of the uppermost Cambrian and lowermost Silurian became a battlefield played out for many years by the two protagonists. On my study wall in front of me, barely two feet from my desk, is a hand-coloured geological map of England and Wales, published by the Society for the Diffusion of Useful Knowledge in 1843. It is both a beautiful work of art and a testament to the ferocious wrangling that occupied the two men for many years. The map was arranged by Murchison, and unsurprisingly when I read the key that explained the colouration, I found that his Silurian was divided into two: the Upper Silurian contained the 'Ludlow rocks and the Wenlock Limestone' found near Dudley in the west Midlands, while the Lower Silurian comprised the 'Caradoc sandstone, Llandeilo Flags & Cambrian Slates'. Sedgwick's Cambrian had been completely annexed by Murchison within eight years of its initial definition! It wasn't until 1879 after the death of the pugilists that this matter was finally settled by Charles Lapworth (1842–1920), the Birmingham professor, who inserted the Ordovician Period between the Cambrian and the Silurian, largely defined on his studies of the planktonic colonial organisms called graptolites that in their fossil form resemble pencil marks drawn on slate.

Before the rows erupted, Sedgwick and Murchison carried out field work in the northern part of Devon, where they mapped red sandstones to which in 1839 they applied the name 'Devonian', naturally enough. It was their final joint geological contribution before they fell out over their Welsh differences, and they went their separate directions to wield their own hammers. It really was most unfortunate, because both were exceptional geologists in their own right, who together could well have unravelled the complex geological history of Wales and other areas besides. The energy they spent sniping at each other could have been used much more productively. Mind you, meetings of the Geological Society would not have been so entertaining! Today marble busts of the two geologists guard the doorway of the Council Room in the Society's London apartments where possibly their ghosts continue to argue and debate the extent of the Cambrian and Silurian to this day.

Murchison provided the name of one other geological Period – the Permian, named after rocks mapped in the Perm region beyond the Ural Mountains in Russia. He had been invited to examine the geology of a large part of northern Russia and so embarked on a 13,000-mile trip in 1841 that took six months to complete. He travelled with the French lawyer and fossil-collector Philippe Édouard Poulletier de Verneuil (1805–1873) and they enjoyed St Petersburg, with visits to its museums and entertainment provided at high society parties and balls. Removing themselves from the capital the geologists headed towards the Urals where Murchison observed some particular successions around Perm, which he referred to as being 'Permian'. The party then moved westward to the Donets Coalfield which they confirmed was Carboniferous in age. In 1845 he published a large two-volume set entitled *The Geology of Russia in Europe and the Ural Mountains*. For his work he had been awarded the Cross of St Ann and was given a beautiful gold and gem-encrusted snuff-box by the Tsar, which can now be seen in the Natural History Museum in London. Soon after the publication of this seminal work, Roderick was to be found at St James' Palace, London, on Wednesday 11 February 1846 on bended knee in front of Queen Victoria receiving a knighthood. He was later appointed Director of the Geological Survey of the United Kingdom, served as President of both the Geological Society and the Royal Geographical Society, and as *The Times* for 13 January 1866 reported, was 'granted the dignity of a Baronet'.

THE ICING ON THE LAYER-CAKE

This summary of the naming of the geological Periods concludes with a brief dip into the background of the terminology of the successions that ice the geological 'layer-cake' – the loose unconsolidated sediments and soil that smother much of the underlying bedrock.

In 1829 Quaternary was used by Jules Pierre François Stanislas Desnoyers (1800–1887) in its modern sense: that is, it included all the unconsolidated material deposited during the last ice age (the Pleistocene Epoch – a term first devised by Lyell in 1839, but restricted by Edward Forbes (1815–1854) to the glacial episode) and during the time following it to the present day (the Holocene Epoch – a term first used by Paul Gervais in 1867, and synonymous with the 'Recent' of Forbes). Desnoyers was a French geologist and historian and a co-founder of the Geological Society of France in 1830.

In 1823 William Buckland had published an important book, *Reliquiae diluvianae*, in which he gave credence to the theory that much of the unconsolidated sediment referred to as 'diluvium' was formed during the flooding associated with Noah. Soon afterwards it was discovered that several different layers of diluvium existed in Europe, but perplexingly none was present in the low and middle latitudes. Buckland changed his tack and with others argued that these various deposits had been formed as a result of several flooding events that were not necessarily global, but were older than Noah's Flood, a concept referred to historians of geology as 'neo-diluvialism'. Buckland was a keen student of subterranean spaces and discovered many examples of cave deposits complete with bones of hyenas, bears and others animals. These deposits he considered to have been emplaced thanks to floodwaters.

Today of course we realise that much of our present-day land-scape has been moulded by two major mechanisms: the action of water and rivers (fluvialism), and the action of ice (glaciation). The dynamic action of rivers was reinforced by Joseph Beete Jukes in a famous paper read in Dublin in 1857 and published later that year in Dublin, where he was able to demonstrate the process of capture by one river of another and the superimposition of a younger southern-flowing river system on an older underlying east–west trending geo-logical structure of mountains and valleys. This produced the unusual right-angled bends on the River Blackwater at Cappoquin, County Waterford, and on the River Lee near Cork in southern Ireland.

We also realise that ice played a dominant role in modifying the landscape. In northern Europe in the 1840s the persuasive Louis Agassiz (1807–1873) championed the glacial cause. Agassiz was at the time best known for his work on fossil fish for which he had received the Wollaston Medal of the Geological Society in 1836. He of course held an advantage over his colleagues who lived in warmer climates, in that he lived in Switzerland and spent a great deal of time studying active glacial features in the Alps around of Neuchâtel. His *coup de grâce* came at the 1840 meeting of the British Association for the Advancement of Science held in Glasgow when he brought to Britain his ideas, and those of others working in Switzerland, on the extent of past glaciations. Ignace Venetz-Sitten (1788–1859), a civil engineer, in a talk delivered in 1821 and published twelve years later, had suggested that the Swiss glaciers formerly extended far beyond their present limits, while his friend Jean de Charpentier (1786–1855), a graduate of Freiberg Mining Academy and a director of the salt mines at Bex (a village situated 26 miles southeast of Lausanne, and once also noted for its sulphur baths), read a paper to the Helvetic Society in 1834 advancing the views of Venetz-Sitten. The reaction in Glasgow to Agassiz was hostile. Following the meeting Agassiz and Buckland embarked on a tour of Scotland, taking in a trip to the parallel roads of Glen Roy, and by the time they reached Charles Lyell at his estate at Kinnordy, Buckland had changed his tune and had embraced the glacial theory. Evaporated from his mind were the neo-diluvial floods. Lyell, too, was a convert, and the three men all presented papers on the topic to the Geological Society in London in November 1840. Later Agassiz was to write of the presentation: 'our meeting on Wednesday passed off very well; none of my facts were disturbed, though Whewell and Murchison attempted an opposition; but as their objections were far-fetched, they did not produce much effect. Dr. Buckland was truly eloquent. He has now full possession of the subject; is, indeed, complete master of it.' However, although the evidence for former glaciations was seemingly incontrovertible, the theory was not generally accepted by most geologists and men of science until the late 1860s.

FURTHER STRATIGRAPHICAL SPATS AND PERSONAL ENMITIES

In the last two hundred years many names have been applied to sections of the geological succession. Some promoted zealously by individuals became accepted, while others disappeared. What is important is that a global standard evolved by which geologists can speak the same stratigraphic language. For the geological Periods this is largely possible.

By and large most of the geological Periods were formulated by the mid-1850s, although conflict and argument between competing geologists tended to cloud the extent, boundaries and status of a number of these, and for these controversial periods later compromise solved the situation. Much has been written about the controversies that affected the delineation of the Cambrian, Silurian and Devonian Periods. Personal feuds and attachment of individuals to particular geographical regions and horizons made rational debate and reasoning difficult. It was not unknown for a particular geologist to latch on to a particular area of ground and to defend that slice of geology and his interpretation of it with the ferocity of a pit-bull terrier. Such jealousies still dog geological research today.

To visit the areas in which the following two examples of geological differences took place, one needs to take a journey beginning in northeast Scotland, continuing through England, and terminating at the southern end of Cardigan Bay in western Wales. In the Elgin region of Scotland, various sandstone units crop out, and the age of these particular units caused some difficulties to workers in the field in the 1850s. Roderick Impey Murchison visited them in the company of a local cleric, the Reverend George Gordon (1801–1893), and declared them to be Old Red Sandstone. This pronouncement was thrown in doubt by the subsequent discovery of reptilian footprints, which suggested a younger New Red Sandstone affinity. Gordon, with another cleric, the Reverend Dr James Maxwell Joass (1829–1914) covered the ground again and confirmed to Murchison that the units could not be separated by breaks in the succession, and they declared the Elgin Sandstones to be Old Red Sandstone. Murchison held to this for some time, but eventually in the fourth edition of his masterful book *Siluria*, published in 1861, recognised that at least some of the units were New Red Sandstone.

St David's is a beautiful part of Wales that is closely associated with that country's patron saint. If one examines a contemporary geological map of the area, it shows that Precambrian rocks are overlain by Cambrian sediments. In 1882 Archibald Geikie, another powerful geologist, who was appointed that year as Director General of the Geological Survey of Great Britain, wrote a long paper on the geology of St David's, which pitted him against the lesser influence of Dr Henry Hicks (1837–1899), a local medical doctor. Hicks was a geologist of some considerable ability in his own right and later served as President of both the Geological Society of London and the Geologists' Association. For some considerable number of years prior to Geikie's interest in St David's, Hicks had carried out investigations of the local geology and had published papers on it in some important geological journals. What aggravated Geikie was that Hicks had described a portion of the St David's succession as being of Precambrian age: in his examination of the geology, Geikie could not agree and placed the succession firmly in the younger Cambrian Period. He presented his paper in two portions to the Geological Society in London at two meetings, on 21 March 1883 and then on

11 April 1883. At that time, before an unfortunate 1970s refurbishment of the meeting room in Burlington House in Piccadilly, the attending Fellows (and non-members) at the meetings of the Geological Society sat in two rows of benches opposite each other – very much in the pattern of a college chapel at Oxford or Cambridge or as in the House of Commons at Westminster. It is likely that Hicks and Geikie sat on opposite benches, which would have added to the dramatic confrontational effect. After the March meeting, Hicks took the opportunity to return to Wales in the company of some students from Cambridge and revisited the areas under question. He held to his reasoning and conclusions, and was present again at the April meeting to hear the second half of Geikie's discourse. Again the paper provoked heated debate and ended with Geikie stating that he hoped that he and Hicks might continue to be friends! The debate was only resolved when the ground was mapped again by Green who published his findings in 1908. While his conclusions were a hybrid of Hicks' and Geikie's conclusions, he did recognise, like Hicks, that the rocks of St David's comprised Precambrian rocks overlain unconformably by Cambrian sediments.

THE COLOUR REPRESENTATION OF GEOLOGY ON MAPS

While the geological column provided stratigraphical firmness, further stability was bestowed on geology when a visual language was adopted. This was in the way that geology was represented on maps.

It is thought that the first representation of geological data on a map dates from 3,000 years ago, when in Egypt some scribe mapped out a district in Upper Egypt on papyrus and highlighted the different rock types present in different colours. This important document is now in a museum in Turin. Today if you walk into any office of your national geological survey you should be able to purchase an attractive wall map showing the geology of the whole country. If you cannot find such a map you should complain! The map will be printed in bright colours, each corresponding to a particular geological Period whose identification is given in a key printed alongside, and it will carry numerous symbols, one of which will show the dip of the rocks, be it at a steep or shallow angle. The standardisation of colours used on geological maps was arrived at quite early, when they were applied as watercolour washes; both Jean-Étienne Guettard's (1715–1786) *Carte minéralogique de France* of 1784 and Walter Stephens (d. 1808) and William Henry Fitton's (1780–1861) 1812 map of Dublin show remarkably close resemblance to maps published only recently. This standard use of colour is most helpful as it allows geologists from any part of the world to interpret instantly the geology of any other part from a map. In essence this colour standard is the Esperanto of geology. All the Silurian strata are identically coloured, as are those of the Jurassic.

Joseph Ellison Portlock, who was responsible for the triangulation of Ireland and who mapped much of mid-Ulster, wrote in 1843 that it was possible to adopt one of two colouring schemes. The first method saw a particular geologically mixed district being given a distinct colour and the mineralogical differences being picked out by the use of symbols. The second method allowed for a different colour to be applied to each particular and distinctive rock type, for example all sandstone being given an identical colour regardless of its age, and the age differences being indicated by symbols. De la Beche used shades of blue to represent limestone of different ages. The second scheme was favoured by most early geological cartographers including those in the survey of Great Britain and in Ireland, although they recognised that some difficulties with the colour pallet had to be overcome. Greenough warned that the colours used should 'speak to the mind as well as to the eye; and the relations of these to one another, must convey clear and definite ideas of the natural relation of the objects which they are employed to represent.' Ingeniously, William Smith represented lithologies on his map in a colour that was as close as possible to the actual colour of the rock itself.

A party of experienced geologists can look at the distribution of the light blue coloration and know instinctively that it represents the

Lower Carboniferous, and if they were to visit a locality in the middle of this blue wash would know it to be underlain by fossiliferous limestone. They know that the brown swathes distinguish the 'Old Red Sandstone' of the Devonian from the pinks of the younger Triassic 'New Red Sandstone'. Equally, they comprehend that the areas of scarlet represent the intrusive bodies of granite that welled up through crustal rocks, and are now only seen at the surface through the erosion of overlying layers of rock. If they were to visit one such mass of red-coloured rock, they would know to bring their walking boots, a stout stick and a good waterproof coat – nine times out of ten they would be entering a mountainous or hilly region.

However, the conventional way of representing geology on maps does present difficulties, and they are frequently not easy for non-geologists to interpret. Modern geological maps illustrate the chronological and not the lithological distribution patterns of rocks. Would it not be better, particularly for those interested in fossil collecting or those engaged in the extraction of stone aggregate for building or road surfacing, to know where the most suitable rock types may be found? Certainly it would. Perhaps the national geological surveys should publish maps showing, for example, the distribution of granite in red, sandstone in brown, siltstone in green, clay in yellow, coal in black, and so on, until all the different lithological rock types were distinguished by a characteristic colour. In reality, however, this would prove almost impossible as there are many different lithologies that could be represented. It could be done if the survey officers printed the maps at very large scales, and used a palate of two hundred colours, but then each map might resemble an impressionist painting, and the information that it was supposed to impart would be lost in the complexity of colour.

By the 1900s the geological column as we now recognise it was generally accepted and important in that it provided a moderately memorable terminology that was recognised globally. Equally the column, and the relative position of any of the geological subdivisions, could be visualised with ease. What was missing, however, at the close of the nineteenth century, was an understanding of the age and duration of any of the geological Periods. The application of an absolute and meaningful timescale to the geological column was undertaken in the early years of the twentieth century.

9 'Formed stones' and their subsequent role in biostratigraphy and evolutionary theory

Fossils are the remains of plants and animals that are preserved in rock. The oldest evidence for life comes from carbon found in rocks on the island of Akilia in southern Greenland that have been dated as being 3,850 million years old. Some authorities have argued that this date is inflated, and that the rocks are in fact 150 to 200 million years younger. Worse still, they have been interpreted as being a banded ironstone, which would indicate that the carbon was not organic. The oldest undisputed fossils are thought to be blue-green algal filaments found in the Apex chert in Western Australia, which are 3,465 million years old. For many millennia, fossils have been the focus of curiosity, but only in the past 150 years have fossils been systematically studied and described. They are useful tools in dating the geological past and have also been used to correlate sequences of rock from area to area, locally or even from continent to continent. The discipline of studying fossils is called palaeontology.

Our understanding of what fossils are and what they can tell us is continually evolving. This is what makes the subject so interesting. Research on fossils is always changing and new methodologies appear every year to test new ideas or re-evaluate old hypotheses. Twenty years ago, it was unheard of to use sophisticated instruments such as mass spectrometers to measure the ratios of the various oxygen isotopes trapped within fossil skeletons, which yield information about the temperature of the sea water from which the skeleton was precipitated. Today, it is commonplace. We are interested in discovering if we can determine how animals reacted in the past to small variations in climate and deduce the implications for present-day life.

In the age of biodiversity studies of latter-day environments, what information can we glean about fossil assemblages and in turn what do they tell us about the nature of past environments that they inhabited? We want to know how complete the actual fossil record is: do the large number of fossil species already recorded over the two hundred years or so since systematic recording began represent almost the full complement of organisms that lived in the past, or a tiny proportion of that past life? Recent studies on a molecular scale have shown what happens to different body parts when death overcomes an organism, and thus shown which parts are more likely to be preserved. This work has helped palaeontologists to understand what the known fossil record is not revealing. It is fully appreciated that soft parts of organisms are rarely preserved, but that this can occasionally occur under exceptional circumstances, such as in the Burgess Shale of Canada and the Soom Shale of South Africa, which have preserved examples of Cambrian and Ordovician biotas respectively. In the past few years there has been a flurry of research on early fossil birds such as *Confuciusornis* from Liaoning Province in China that has revealed clues to avian ancestry or phylogeny. By focusing on the lithological and environmental characteristiscs of these exceptional fossil formations, or *lagerstätten* as they are generally known, geologists have been able to home in on other sites in which similar occurrences of fossils may be found.

In the 1950s and 1960s palaeontologists began to appreciate that studies could be enhanced by looking closely at the morphology of fossils and deducing what soft parts had been present when the organism had been alive. From this the function of elements could be determined and this in turn led to a greater understanding of how and where the organisms lived. Equally interesting were the studies in distribution of species, research driven by the newly garnered evidence of plate tectonics. Palaeontologists could plug in the data they had relating to the stratigraphical and geographical distribution of fossils, and use the information on where the lithospheric plates had been in the past to get some idea of the palaeobiogeography of the past floras and faunas.

In 1938 a large fish was dredged up by Hendrik Goosen, a local fishermen, just off the coast of East London in South Africa. He had never seen such a fish before, and so gave it to the local museum, whose curator Marjorie Courtenay-Latimer (b. 1907) was also puzzled by its appearance. She in turn contacted Professor James Leonard Brierley Smith (1897–1968) of Rhodes University in Grahamstown, who recognised the fish as being a coelacanth, a form that had previously been known only from fossil examples and was thought to have long been extinct. He announced the find in a short paper published in *Nature* in 1939 and it caused a sensation. Further examples were later recovered, particularly from around the Comoros Islands, which lie off the coast of northern Mozambique. Here was a 'living fossil' which could be used to reinterpret the functional morphology and live habits of its similar fossil sisters and brothers. Examples of living fossils are rare: *Lingula*, a simple Cambrian inarticulate brachiopod, has remained almost unaltered for just over 500 million years; *Nautilus*, the pearly cephalopod that lives in the Indian Ocean, is a modern analogue for the ammonites, a major Mesozoic group that became extinct 65 million years ago.

In the nineteenth century much palaeontological work was concerned with the description and illustration of fossil genera and species, and numerous monographs appeared. This was really rather mundane stuff, but it remains important to the present day as these monographic studies provide a primary database on which many modern studies draw. They were also invaluable as they provided greater precision to the biological-based stratigraphy, a scheme which William Smith had initiated in 1816. And they were vital to the discipline of biostratigraphy, the methodology of subdividing and correlating the geological time periods on the basis of their characteristic fossils. However, let us leave William Smith for a moment and consider an important issue. Fossils have not always been recognised for what they are, and in the past numerous explanations have been offered for their formation and nature.

EARLY NOTIONS ON FOSSILS

As noted in an earlier chapter, both da Vinci and Steno recognised that the fossils that they had found in the mountainous regions in Italy were organic in origin. But this understanding was not always evident. Fossils have fascinated people for at least six millennia, and in order to understand them, people came up with a host of explanations for their origins and for their functions. Most of these could be described as belonging to the realm of 'folklore'.

Fossils were noticed and collected by Neolithic man and placed inside burial tombs. About eight years ago a diverse assemblage of Lower Carboniferous fossils was located inside a 4,500–5,000-year-old passage tomb in southwest Ireland, and this occurrence pre-dates similar instances in Bronze Age burial chambers in Britain by 500 to 2,000 years. The fossils include brachiopods, gastropods and cephalopods and were collected from the immediate area and placed within the tomb by its builders, who regarded them as holding some ceremonial or decorative significance.

Native Americans made pendants and necklaces of various fossils for adorning themselves. It is unlikely that these groups of early fossil users considered where the material had come from, and they probably did not make the connection between fossils and living organisms. The Greeks had worked this out by the sixth century BC, and it is probable that the Chinese had done likewise. When Pythagoras (540–510 BC) noticed shells in mountainous rocks he deduced that the mountains must have been below sea level at some time in the past. Pliny (the Elder), who lived between AD 23 and AD 79, noted several fossils, including shells and sponges, in his writings, and correctly attributed amber to pine trees. In his *Natural History* Pliny described snake eggs that were reputed to be strong antidotes against snake poison. These fossils were actually sea urchins. He also described some fossils as tongue-stones (*Glossopetrae*), which were correctly described some sixteen centuries later as sharks' teeth by the Italian Fabio Colonna (1567–1650).

Throughout mediaeval times the true nature of fossils seems to have been lost to researchers. The prevailing view was that fossils originated in the bowels of the Earth from a creative or plastic force (*vis plastica*). One of the first books in the English language in which fossils were described was *The Natural History of Oxfordshire* (1676) by Robert Plot whom we have met in Chapter 4, but he collected them together under the heading 'Formed Stones'. While the notion of a plastic force was common so too was that of a universal Flood. This idea began to subside as a plausible cause for the formation of fossils by the middle of the eighteenth century. Many enlightened thinkers throughout the Renaissance, and after, were unconvinced of the organic nature of fossils, although some were certain that fossils represented the remains of past life.

Among popular folklore, many fossils have been given names that allude to their supposed origin. The cone-shaped shells of the cephalopod belemnites were thought to be petrified thunderbolts on account of their pointed shape; the bivalve *Gryphaea arcuata*, commonly found in rocks around the River Severn, was called the Devil's toenail; the echinoid *Micraster* has a heart-shaped test which was known to local people on the south coast of England as a fairy loaf; and the Silurian trilobite *Calymene blumenbachii* from the Silurian of the Dudley region was called the Dudley locust, and actually appeared on the coat of arms of that town. Another case of fossils being used as urban symbols was that of Whitby, Yorkshire, which proudly displayed snakestones. These are actually ammonites which have been carved with heads by local people (Figure 9.1).

The Englishman Martin Lister (1638–1711) produced some beautiful illustrations of fossils in his book *Historia animalium Angliae* (1678). However, he did not regard fossils as being the remains of animals and supposed them to be imitations caused by unknown forces. Shortly afterwards, in 1699, Edward Lhwyd produced the first book exclusively related to British fossils. This was *Lithophylacii Britannici Ichnographia* in which he illustrated over 200 fossil species.

Figure 9.1   Whitby snakestones (the largest has a diameter of 55 mm). These are actually Jurassic ammonites *Dactylioceras commune* onto which a head has been carved. (Geological Museum, Trinity College Dublin.)

BEGINNING OF SCIENTIFIC INVESTIGATION

The binominal system of naming organisms which was devised by the Swedish naturalist Carl Linnaeus in the eighteenth century began to be widely used. From then on all living and fossil organisms were given a genus name, e.g. *Homo*, and an epithet for the species, e.g. *sapiens*. By the early 1800s most scientists were convinced of the organic nature of fossils, and researchers began to describe and illustrate fossils in a systematic manner. While many of the early descriptions of this period were short, the illustrations were generally of a high standard. Two early attempts at the beginning of the nineteenth century at a systematic treatment of fossils merit mention: John Parkinson's (1755–1824) *Organic Remains* (1804), and William Martin's (1767–1810) *Petrificata derbiensia* (1809).

THE BEGINNINGS OF BIOSTRATIGRAPHY

By the middle of the nineteenth century the scientific community had refuted the more fabulous hypotheses suggested for the origin of

fossils, and it is reasonable to say that by then the science of palaeontology had begun. In France, Georges Baron Cuvier championed
palaeontological research and produced many important papers
including his celebrated *Description géologique des environs de
Paris*, co-authored with Alexandre Brongniart (1770–1847) in 1822.
In it they described the Tertiary deposits of the Paris Basin, and, as
Martin Rudwick pointed out in 1997, used the fossils as indicators of
palaeoecological conditions rather than strict stratigraphical tools as
William Smith had. In his *Strata Identified by Organized Fossils,
Containing Prints on Coloured Paper of the Most Characteristic
Specimens in Each Stratum*, published in four parts (of a projected
seven) between 1816 and 1819, and *Stratigraphical System of
Organized Fossils*, which appeared in 1817, Smith recognised two
fundamental principles: that successive rock sequences were characterised by differing fossil assemblages, and that a stratum found
beneath another is younger unless it is shown to have been overturned. These principles allowed for the easy determination of the
relative age or position of strata through an inspection of its contained
fossils. They paved the way for the development of biostratigraphy –
the study of rock sequences based on the fossils that they contain.

In *Strata Identified by Organized Fossils*, each part (priced at
7 shillings and 6 pence) consisted of a number of pages of explanatory
text that preceded between three and five plates. Each plate depicted a
series of fossils from a particular horizon printed on a background
colour that represented the colour of the lithology in which they were
found (Figure 9.2). This was an ingenious method of palaeontological
colour coding, and the colour coincided with those used on his great
map of 1815. The accompanying text for each illustration gives the
name of the horizon such as the 'Upper Chalk' or 'Fuller's Earth
Rock'; the nature of soil that developed on top of it; notes on the purity
or otherwise of groundwater found associated with it; and it lists the
illustrated fossils and the localities from where they were collected. He
noted that Suffolk contained some of the worst land in the country on
account of the presence of blown sand, and that shelly deposits occurred

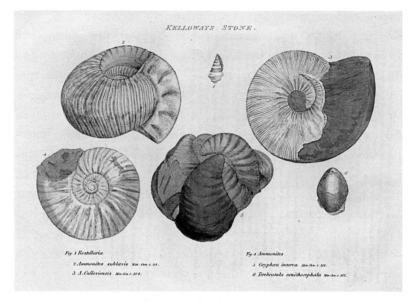

Figure 9.2   Plate illustrating fossils from the Jurassic Kelloways Stone from William Smith's *Strata Identified by Organized Fossils* (1817). 1 is a small gastropod; 2–4, ammonites; 5, the distinctive bivalve *Gryphaea* (colloquially known as the Devil's Toenail); and 6, a brachiopod. Courtesy of Hugh Torrens.

in a blue clay found in Essex that produced a 'tenacious soil'. Smith confidently proclaimed the ease with which his scheme could be used: 'The organized Fossils (which might be called the antiquities of nature) and their localities also, may be understood by all, even the most illiterate: for they are so fixed in the earth as not to be mistaken or misplaced ... consequently, they furnish the best of all clues to a knowledge of the Soil and Substrata.' In *Stratigraphical System of Organized Fossils* Smith described in detail his fossil collection, which because of debt he was forced to sell to the British Museum in batches between 1815 and 1818. The money raised was not enough to save him from the debtors' jail and he spent ten weeks incarcerated in the middle of 1819. Smith planned to issue *Stratigraphical System* in several parts, but in the end only the first part comprising 119 pages appeared.

As a spin-off from his geological map of 1815 Smith also produced a *Geological Table of British Organized Fossils* that took the form of a single sheet of paper approximately 17 inches by 20 inches in size. This listed in order thirty-four stratigraphical units and gave the colours with which they were represented on his map. Information on their mineralogical constituents and the characteristic fossils contained within them were given in two additional columns. Soon similar tabulations of fossils appeared from the pen of other authors. Among the most attractive was that published in 1853 by the geological and cartographic publisher Edward Stanford of Charing Cross Road in London. This took the form of four fold-out charts mounted on linen on which the geological time divisions were marked on the left-hand side, beside which the major lithological formations were given, together with their thicknesses and physical and palaeontological characteristics. The remaining three-quarters of the chart was given over to drawings of fossils by the noted geological illustrator and map colourist Charles Richard Bone (1809–1875) who worked for the Geological Survey. The drawings were engraved and compiled by James Wilson Lowry (1803–1879) whose sister Delvalle was the author of several geological texts. By unbinding the four charts from the confines of their flimsy cover the user could mount them in a poster arrangement and hang them on their office or laboratory wall to produce a easily consulted visual guide to the palaeontological succession of Britain. Similar charts on a smaller scale were made available in many general texts. Edward Clodd (1840–1930) in the revised 1896 edition of his book *The Story of Creation: A Plain Account of Evolution*, a book that proved to be highly popular and helped to disseminate Charles Darwin's (1809–1882) theory of evolution to non-scientific audiences, includes such a chart that shows a small image of *Eozoon canadense*, 'the Dawn Fossil', propping up the base of the geological column, with the Irish Elk at the top representing the Recent (Figure 9.3). Interestingly, both of these fossils proved to be problematic. *Eozoon*, which was considered to be the earliest known example of life on Earth, became the focus of a major palaeontological debate in the late 1800s and was later shown to be of inorganic origin, while the 'Irish Elk' is actually a member of the

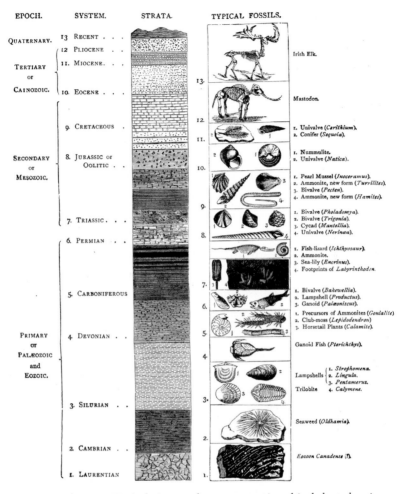

Figure 9.3   Typical nineteenth-century stratigraphical chart showing diagnostic fossils for each geological period (from E. Clodd, *Story of Creation* (1896), facing p. 32).

deer family and is not related at all to the familiar North American animal.

ACQUISITION OF FURTHER BIOSTRATIGRAPHICAL DATA

From the middle of the nineteenth century many papers that documented the variety and distribution of fossils had already appeared in the major geological journals such as *Annals and Magazine of Natural History* and Benjamin Silliman's *American Journal of Science*, but it was not until the middle of the twentieth century that more specialist palaeontological societies and journals began to emerge. In the 1800s a major vehicle for the publication of palaeontological information was the monograph. In 1847 the Palaeontographical Society was established purely to publish such volumes on British fossils. A century and a half later it is still active, and during that time has produced numerous volumes that would fill nearly five metres of shelving. It has published works by the leading palaeontological authorities, on diverse subjects including Mesozoic reptiles, Eocene fish otoliths and Cambrian trilobites. Richard Owen (1804–1892), the founder of the Natural History Museum in London, produced seven monographs, one of which was on the reptiles of the Wealden and Purbeck formations. Published in parts, which was typical, the first appeared in 1853 and the final part and title page in 1889 three years before his death. Among other classic titles produced by the Society were those by Thomas Davidson (1817–1885) on brachiopods (in six volumes, 1851–1886), Charles Darwin on fossil cirripedes (1851–1855), and more recently the separate monographs on ammonites by Sydney Savory Buckman (1860–1929), and William Joscelyn Arkell (1904–1958). In Ireland the most important paloeontological monograph was Frederick M'Coy's (1817–1899) *Synopsis of the Characters of the Carboniferous Limestone Fossils of Ireland* which was published privately in 1844 by Richard Griffith. In it a wide variety of taxonomic groups of Irish fossils was described, many of which were new genera or species. Like many collections dating from the mid-1800s the specimens described by M'Coy are still to be found in museums and individual specimens can be recognised from his illustrations. In Belgium, Laurent Guillaume de Koninck (1809–1887) published an important series of monographs on fossils of a similar age, while Joachim Barrande (1799–1852) described in twenty-two volumes the Lower Palaeozoic fossils in his own cabinet, which he had collected from around Prague; this region is now called the Barrandian and is considered to be one of the most important globally for the study of Silurian rocks. Alicide Dessalines d'Orbigny (1802–1857), a remarkable French naturalist and traveller, published during his short life a number of very important works on Mesozoic and Cenozoic fossils, of which in the present context his *Prodrome de paléontologie stratigraphique universelle* (1850–1852) was the most valuable. D'Orbigny highlighted some difficult fossil groups and he paid particular attention to microscopic organisms including Foraminifera, unicellular animals that continue to inhabit the world's oceans. It was later shown that these microfossils had great biostratigraphical potential. In the United States important monographic series included *Palaeontographica Americana* published since 1916 by the Paleontological Research Institute of New York, and in Germany *Palaeontographica*, which commenced publication in 1846.

Students in the mid to late 1800s had access to a plethora of palaeontological information, and could either consult the primary monographs or resort to a new source of secondary material, the palaeontological textbooks. Of these the *Traité élémentaire de paléontologie* (1844–1846) by François Jules Pictet (1809–1872), the *Manual of Palaeontology* by Henry Alleyne Nicholson (1844–1899) which first appeared in 1872, and the *Grundzüge der Palaeontologie* by Karl Alfred von Zittel were probably the most comprehensive and therefore the most useful. The German text was first published in 1895 in Munich, and an English edition (*Text-book of Palaeontology*) followed in 1902 and remained in print for at least thirty years. Geologists were now armed with the raw ingredients with which to apply biostratigraphical analysis.

MARKER FOSSILS, OR BIOZONES

Smith developed a crude biostratigraphy, but thanks to the numerous studies in the thirty years following his work biostratigraphical precision was tightened up considerably. In Germany two palaeontologists made the study of Jurassic ammonites their lives' work. Ammonites are extinct molluscs that are members of the Class Cephalopoda, a classification they shared with squids and cuttlefish. They lived between late Devonian and Cretaceous times, and reached their greatest diversity and success in the Mesozoic. Friedrich Quenstedt (1809–1889) was a professor of palaeontology at the University of Tübingen and he interested a young student, Albert Oppel (1831–1865), in these fascinating animals. Oppel was not to live long, but in his short life he overshadowed Quenstedt because he succeeded in demonstrating that the Jurassic of southern Germany contained many ammonites which, crucially, appeared to have a short range. He mapped out the distribution of fossils in particular beds and recognised that narrow horizons could be characterised by overlapping ranges of ammonite species. These Concurrent Range Biozones are now known as 'Oppel Zones' and the fossils that characterise particular biozones are known as index fossils. Ammonites provide a zonal precision of as little as 200,000 years. In other cases the range of a single species can provide a Total Range Biozone, while evolutionary sequences of fossils can be used to define a Consecutive Range Biozone where one particular species rapidly evolved into another species which in turn was succeeded by another. Other fossil groups are also useful in zonation, and trilobites and graptolites have been used as index fossils for the Cambrian and Ordovician respectively. The most useful zone fossils are those that had a global distribution and a rapid evolutionary turnover, and that are commonly found.

Soon after Oppel's work, zonations for many areas began to appear in the geological literature, and palaeontologists expended a lot of effort in determining the ranges of fossils. They wanted to know when each type first appeared and when they disappeared, to ascertain whether a particular species had biostratigraphical potential. In 1909 and 1910 Amadeus William Grabau (1870–1946) and Hervey Woodburn Shimer (1872–1965) published their landmark two-volume *North American Index Fossils* that provided an illustrated condensation of all the systematic works published for that continent and illustrated the most important fossils that stratigraphers would wish to identify during the course of their attempts to correlate and date sequences of fossiliferous rocks. The volume was reissued in 1944 in a revised form by Shimer and Robert Rakes Shrock (b. 1904). While it is still of some use, perhaps of greatest value to geologists today is the multi-volume work *Treatise on Invertebrate Paleontology* which has been published since 1953 by the Geological Society of America and the University of Kansas Press. Conceived by the stratigrapher and palaeontologist Raymond C. Moore (1892–1974) it provides separate volumes for different biological groups. Each genus is described and illustrated and their age range is given. It is the foremost identification aid available.

CORRELATION CHARTS

Today many geologists who study particular periods in geological time do not confine themselves to limited geographical regions, but attempt to learn what was happening geologically in other more widely flung districts worldwide. To do this they have to rely on correlation charts. One of the earliest such documents was that arranged by the English geologist Bernard Hobson (1860–1933), a lecturer in petrology at Owens College, Manchester, whose charts provided data for Britain and Ireland. In the late 1960s the Geological Society of London took it upon itself to establish a number of committees each given the task of producing detailed correlation charts and an accompanying explanatory volume of British and Irish strata in one of the geological Periods. By the late 1970s most Periods were covered and in fact in the past ten years a number of reports have been revised in the light of new findings. Similarly correlation charts for North America have been published by the United States Geological Survey and by the Geological Survey of Canada, and many other areas have these aids to correlation. On a local scale correlation charts are easier to produce than those on a global scale. For some geological Periods global standards exist, for others they are piecemeal and global standards are in the process of being formulated and agreed. Such agreements may take decades.

Another problem facing compilers of correlation charts is that of terminology. Many names have been used for geological horizons that are now known to be coeval or equivalent. In the Harlech Dome in Wales the Cambrian Garth Hill Beds crop out, and these are correlated with the uppermost Lonan and Niarbyl Flags of the Isle of Man. Fortunately this problem was addressed by a commission established by the International Geological Congress and it arranged for the publication of the *Lexique Stratigraphique International*, which provided information on geological names used in nearly every country in the world; this resource has proved to be of huge value to stratigraphers attempting tricky intercountry correlations.

EVOLUTIONARY QUESTIONS AND THE PAUCITY OF TIME

It was inevitable that at some point in the past an interested scientist would ask questions pertaining to the ancestry of fossil groups and the sudden appearance and disappearance in the fossil record of certain plants and animals. Geologists and biologists recognised that strata of a particular age contained fossils of organisms that appeared to be no longer extant. What had happened to them? Cuvier in 1801 had asked this question and speculated that animals might have become extinct, might have evolved into another species, or might have migrated to another place. He favoured extinction. Early workers also realised that some organisms appeared to have linkages, such as the animals with backbones, the chordates. Here the first fishes were found in the oldest successions and preceded the amphibians, which in turn were followed by reptiles and finally the mammals which were confined to the younger sediments. Were these groups interrelated, and if so, had shared morphological characters been passed on? Could supposed shared features appear independently in different groups? Is the fish fin related in any way to a human humerus? Could one organism metamorphose into another similar organism?

These questions, which we would now group into the realm of evolution, were first tackled in the early 1800s. Jean Baptiste Pierre Antoine de Monet de Lamarck (1744–1829) was a soldier until he was wounded in 1726, which resulted in his being invalided out of the French army. He then turned to the study of natural history and joined the staff of the Muséum National d'Histoire Naturelle in Paris. He produced numerous books, on the flora of France and hydrogeology amongst other subjects, and he is today chiefly remembered for his ideas in zoology and the inheritance of characteristics from generation to generation. Martin Rudwick, writing in 1985, said of Lamarck's ideas that 'In his great *System of Invertebrate Animals* (1801) he asserted that "one must believe that every living thing whatsoever must change insensibly in its organisation and in its form"; and given enough time such slow changes were not the least surprising, he argued, for there had been ample time for the fossil forms to turn into the living ...'. Lamarck went on to say that 'one may not assume that any species has been really lost or rendered extinct.' He was saying that species were transformed into new species, as against Cuvier's views that species could become extinct. Lamarck was on unstable ground in that he had little understanding of the elements of stratigraphy and was not an experienced palaeontologist. He was unlike Cuvier, who although younger had already begun his rigorous work on the fossil faunas of the Paris Basin which was enhanced by anatomical comparisons with living organisms. By the end of his life Cuvier was considered to be the doyen of French palaeontology and had been the recipient of numerous honours.

In 1844 Robert Chambers (1802–1871) published *The Vestiges of the Natural History of Creation* that questioned whether the sequence of fossils seen in the geological record represented transformed species of those originally created by a divine hand. Tampering with God's plans did not go down well in intellectual circles and it was just as well for Chambers that he published his book anonymously.

Fifty-eight years after the publication of Lamarck's *Système des animaux sans vertèbres* (1801) a book was published in England that formed the basis of our present understanding of evolution and its mechanisms. Its author was Charles Darwin, who had spent five years circumnavigating the globe in HMS *Beagle* on its voyage between December 1831 and October 1836. Darwin served as the ship's naturalist and while the study of geology was his first love he assembled plenty of biological evidence that he later used in the ground-breaking book *On the Origin of Species by Means of Natural Selection*, which appeared in the booksellers on 22 November 1859. It would not be a wild speculation to suggest that more copies of the Bible have been circulated than have any other printed book, but it is probable that *Origin of Species* would occupy a place in the top ten all-time bestsellers.

Following his return to England in 1836, Darwin wrote a number of geological books and began to ruminate on biological ideas. Within three years he had formulated the kernel of his theory, but he lacked the confidence to commit his ideas to print. He was only forced to do so over twenty years later when Alfred Russel Wallace (1823–1914) wrote to him in June 1858 concerning his own ideas on natural selection and the formation of species. Darwin realised that both men shared a similar vision and so together they wrote a paper that was published by the Linnean Society in 1859 in London.

In *Origin of Species* Darwin proposed that new species emerged as a result of modifications to earlier species, but that the older species did not necessarily disappear. He argued that variation could occur spontaneously, and that when the new species was better suited to its environment than earlier species, natural selection had occurred. He recognised that geographical isolation could drive speciation, and he became interested in the mechanisms and timing of agricultural and horticultural breeding that produced new varieties.

His conclusions drew a very mixed reaction, with many in the Church and the scientific community outraged by what he had written. On the other hand, he also had some supporters. In terms of geological time and the age of the Earth, Darwin's evolutionary ideas created a headache for many. If the elements of the fossil record were not created by God, and were the result of progressive changes through time, it was clear that in order to produce the great diversity of plants and animals known from the fossil record in the way that he outlined then an enormous span of time would be required. This time frame, his opponents argued, was quite unrealistic.

10 The hour-glass of accumulated or denuded sediments

In July 1798 Napoleon Bonaparte's army invaded Egypt and commenced a huge and famous survey of its antiquities and natural history. The French hold on the region was weakened following the celebrated Battle of the Nile when Horatio Nelson (1758–1805) showed his prowess, and the British assumed control in 1801. Thereafter the fashion for things Egyptian spread to England and remained much in vogue until the 1920s. There is a nineteenth-century children's ditty that describes the foolish woman who, grinning to her worried friends, set off to explore the River Nile riding on the back of a crocodile: 'at the end of the ride the lady was inside and the smile was on the crocodile.' An earlier traveller was more fortunate, and he lived to make one of the earliest contributions to geological literature.

Herodotus (484–408 BC) was a Greek traveller born in Western Anatolia (what is now Turkey) who has been styled the 'Father of Geography' on account of his writings and observations on the changes effected on the Earth's surface by river action and erosion. Although Herodotus did not travel widely by modern standards (his world was a triangle drawn between Greece, Italy and Egypt), by standards in the fifth century BC he would have clocked up plenty of 'mileage points'. On one trip he sailed to Egypt and onwards up the Nile, preferring to use a boat unlike the crocodile-riding lady, and the first thing that struck him as he approached Egypt was that the sea was very shallow far out to sea, and that it continued to shallow as he approached the delta front a day's sailing away. Once he arrived at the delta, he noted how flat the land either side of the river channels was, and how fertile it was. He also appreciated that the sediment had been carried a great distance by the river. Herodotus was one of the earliest thinkers to attempt to quantify sedimentation rates. He calculated that it would take 5,000 years for the Red Sea to silt up completely. This sedimentation theme was taken up nearly 2,400 years later in England and still later in the United States, and for a while occupied the thoughts of geochronologers until it was swept away by the presentation of a new theory.

SEDIMENTATION RATES AND GEOCHRONOLOGY

The basic premise underlying the use of sedimentation rates as a tool for estimating the age of the Earth is that if one can estimate the thickness of a modern sedimentary deposit such as a delta, and one knows the rate at which sediment was added to it over a period of a year, then a simple mathematical calculation will give the length of time that the delta has been forming. Similarly, if one knows the original height of a feature on the surface, such as the Colorado Plateau, and one measures the depth of the Grand Canyon and knows the rate of downcutting by the Colorado River into the level sediments, one can calculate the length of time that has elapsed since the beginning of canyon formation. Simple, it would seem, but in actuality this calculation is not so, even for the seemingly straightforward deltaic model. It is even more complicated when you look at the thickness of the sedimentary rocks that make up much of the geological record. In an ideal situation (but the Earth is never that helpful), it suffices to say that if we know the thickness of the sedimentary pile that makes up the complete rock succession and the rate at which it was deposited we should be able to estimate the age of the Earth.

Denudation, or the breakdown of rocks exposed at the Earth's surface, leads to the topography seen today. Denudation produces two different types of product: firstly, particulate matter or 'clasts' of a variety of sizes that range from cobbles (the largest grains) through sand and silt, to mud (the finest fraction), and secondly, unseen ions in solution in water. Both products were used as geochronometers – as we saw earlier, Edmond Halley discussed how an estimation of the rate at which a freshwater body became salty might be useful in estimating the passage of time, and this is a theme that we will revisit later. The use of fragments of rock, or sediment, is discussed here.

DELTAS

The fourth letter of the Greek alphabet, Δ, is applied to triangular deposits of sediment found occasionally in lakes but more usually at the mouths of many rivers where they drain into the sea. These features may be of varied types such as arcuate (the Nile, Rhône and Po), bird's foot (Mississippi) and cuspate (Tiber), and may be complex, comprising more than one river channel and many distributaries each divided from its neighbour by sand bars and low flat islands. In addition deltas may consist of many gradually moving complexes: sixteen lobes have been mapped out in the Mississippi delta, the oldest of which is 7,000 years old. The best-known deltas in the world are those of the rivers Ganges and Brahmaputra in northeast India (which form one system and are generally treated together) which has an area of 105,000 square kilometres, the River Mississippi that flows into the northern shore of the Gulf of Mexico, the Amazon that feeds sediment into the western mid-Atlantic and the River Nile that empties into the Mediterranean. The River Amazon disgorges 1.2 thousand million cubic metres of predominantly suspended silt and mud every year, and as a result the delta front is prograding, or growing out into the ocean at a rate of 5–10 cm a year. The Ganges is not the longest river in the world (that honour belongs to the Nile at over 6,600 kilometres) but it deposits more material than does its African counterpart. This is because each river system is unique and its physical features and the volume of material that it carries are controlled by a number of factors such as the river profile, which broadly is the slope of the river from source to estuary. The greater the slope, the greater the power of the water carried within the river to erode the terrain through which it flows. The Nile flows over largely flat expanses of the north African continent, in direct contrast to the Ganges which drains the Himalayas, which are continuing to rise following the collision of India with the Asian continent some 50 million years ago. This meeting of two continental masses led to the development of the highest mountain chain on Earth and the Tibetan Plateau behind, and today erosion rates in the peaks are high. Erosion and uplift, by and large, keep in equilibrium so that the mountains in Nepal do not appear to be getting higher. Another factor that controls sediment load in a river system is the size of the catchment area drained by the river and all its tributaries: simply said, the larger the catchment, the greater the load potential – that of the Mississippi is 3 million square kilometres, larger in area than the Ganges and Brahmaputra system which itself is just over a million square kilometres. Another factor is the discharge rate of the river: how much water does it carry? If it is sluggish and meanders slowly and is usually shallow, then less sediment will be carried than if the river is speedy and deep. Finally, the sediment load carried annually will frequently be greater if the flow regime of the river is constant throughout the year.

UPLIFT AND DENUDATION: MECHANISMS OF SEDIMENT PRODUCTION

The ability of a river to erode the rock over which it flows will be controlled in turn by a number of physical and dynamic states, of which the following two deserve particular mention: the uplift of the continental masses, and the nature of the rock over which the river-water flows.

Uplift of rock masses is caused in various ways and often at very different scales, from movement affecting whole continents to local movement of a number of millimetres along a fault. It is now recognised that the continental masses, which are generally granitic in composition, sit or 'float' on a bed of denser darker basaltic material that is also found beneath the oceans. This underpins the concept of 'isostasy', an idea devised and explained in two models by John Henry Pratt (1809–1871), an Anglican cleric who ministered in India, and by Sir George Biddell Airy (1801–1892). During the last ice age much of northern Europe was blanketed with ice-sheets up to 1 kilometre in thickness. This extra weight depressed the northern margin of the continent and so when the ice finally melted the area experienced isostatic rebound. Today the northern part of Norway continues to rise at a greater rate than southern Norway. This naturally imposes differences on the dynamics and flow regimes of rivers in the two ends of the country. This change in Norwegian elevation is rather local. If two continental masses collide, as happened 450 million years ago when the Iapetus Ocean closed and a land mass broadly coincidental with North America of today crashed slowly into a land mass that makes up modern western Europe, the result is uplift on a large scale. Mountains with the altitude of the modern-day Mount Everest were thrown up in Scotland and northwest Ireland and the area was dotted with volcanic activity. The pimples that make up the Scottish peaks collectively known as Munros are the eroded remnants of these great mountains. Similar collisions have produced the Appalachian Mountains and the Alps.

Uplift is sometimes associated with volcanic activity, and perhaps the classic case of this is the Andean Belt that was produced as the Pacific Plate was subducted beneath the South American Plate. As the oceanic rocks of the Pacific Plate were pushed deeper, they melted, and the molten rock found its way towards the surface where it was extruded in volcanoes such as andesite and rhyolite. Many of these volcanic slopes are highly unstable and have yielded copious quantities of sediment in a relatively short period. Fifty million years ago, as the Atlantic Ocean was opening and North America was moving away from Europe, hot plumes of volcanic material centred on the area around Mull and Ardnamurchan on the western side of Scotland led to emplacement of volcanic rocks and associated uplift. This uplift affected the flow regimes of rivers in the area and the erosion of the pre-Tertiary rocks by those rivers. As has been recently demonstrated by various researchers interested in the denudation of northeast Ireland, it is not sufficient to examine the flow patterns of extant rivers in order to achieve an understanding of how the ancient rivers behaved.

Not all rocks are the same and it is rare to find a river catchment area draining only one type of lithology, and equally unlikely that one will find two different rivers flowing over identical rock types in equal proportions along their length. Some rocks are harder than others – the rocks that make up the central and ancient 'cores' of continents tend to be crystalline igneous rocks such as granite and granodiorite, and metamorphic rocks such as schist and gneiss, and these are far more resistant to erosion than the usually younger overlying sedimentary rocks. Inequality of strength is also displayed in similar sedimentary rocks; not all sandstones have the same physical characteristics although all are composed of cemented grains of quartz (silicon dioxide), which itself is a hard mineral. On the mineralogical scale of hardness, which was devised by the German mineralogist Frederick Mohs (1773–1839) and carries his name, quartz is ranked 7 (diamond, the hardest substance, is 10, while calcite or calcium carbonate which makes up most limestones is 3). The strength of a sandstone or indeed of any rock is only as strong as its weakest component; in the case of sandstone, many are cemented by calcium carbonate or calcite, and so in reality are more susceptible to erosion than those cemented with silica.

Close examination of a sediment sequence or a sedimentary rock will reveal information about sediment type, provenance, size and sorting and this can be used to interpret something about the environment in which the material was deposited and the mechanism that produced the sediment, but often it is difficult to gauge the speed or timing of deposition. In the past the rates of denudation were not always uniform and varied from location to location and environment to environment.

THE SEDIMENTARY HOUR-GLASS

How did geologists use the sedimentary record as an hour-glass? By the 1850s the geological community had taken on board the ideas of James Hutton and accepted that the Earth's history was cyclical, but in the context of sediment accumulation the uniformitarianism ideas spread by Charles Lyell were more important. If the various active geological processes of the past were similar to those active at the present time,

Figure 10.1 John Phillips
(1800–1874) (from *Geological
Magazine* 7 (1870), facing p. 301).

then the characteristics of the present environments and the rocks
that they produced could be used to get a portrait of what the past was
like. The geologists were familiar with a large range of sedimentary
rocks and would have been pretty confident that they knew how and
where they formed. In 1860 the English geologist John Phillips
(Figure 10.1) first estimated the actual age of the Earth using sediment
accumulation as a indicator of time. From the 1830s, he had recognised that its antiquity was of enormous duration although he did not
attempt to quantify quite how enormous. Following his 1860 calculations, others in Britain and Ireland and in the United States followed
suit using his methodology.

Given that Phillips was orphaned at an early age and his formal
education, which was undertaken in Wiltshire, ceased at the age of
fifteen, he had an extraordinarily successful geological career: he held
chairs in London at King's College (1834–1839), at Trinity College
Dublin (1843–1844), and at Oxford (1860 until his death in 1874).
Born on Christmas Day 1800 at Marden in Wiltshire, he lost both of
his parents before his eighth birthday, and he was raised by his uncle

William. Probably much of his later geological success was owed to his uncle's influence; this was none other than William Smith, the so-called 'Father of English Geology' whose stratigraphical laws and mapping did much to regularise the vast lithological record (Phillips, it will be remembered, made his own stratigraphical contribution when he coined the stratigraphical terms 'Mesozoic' and 'Kainozoic' in 1840). From 1815 Phillips trained as a surveyor and assisted his uncle's various geological enterprises. Unfortunately Smith lost his London base when he was imprisoned for bankruptcy and following his release he and his nephew headed northwards to Yorkshire where they eked out a peripatetic existence. In 1825 Phillips was appointed Keeper of the Yorkshire Philosophical Association's museum and his geological career began in earnest. He too, like his uncle, produced a large-scale geological map, but expanded the coverage to include the whole of the British Isles, and later adjacent areas of France. The first edition appeared in about 1837 with the eleventh and final addition twenty-five years later in 1862. Unlike his uncle, Phillips worked for the Geological Survey and had hoped to be appointed Local Director for Ireland, a position he had aspirations to hold concurrently with his Dublin chair, but (probably through geological infighting) this avenue was closed to him and he returned to England.

Phillips was closely associated with the British Association for the Advancement of Science and was instrumental in the early organisation of its important meetings, which migrated from city to city on an annual basis. He was to act as the Assistant Secretary to this organisation for over thirty years. The inaugural session was held in 1831 in York where Phillips was working at the time. During his residency in York, Phillips also produced a classic account of the geology of the region in two volumes. Subsequently his geological collection on which he based some of his findings given in these volumes ended up in the collections of William Gilbertson and a number of these, including the type of the important bryozoan *Fistulipora*, are now in the Natural History Museum, London. He was appointed a deputy reader in geology at Oxford in 1853, reader in

1856 and professor in 1860. Concurrently he held the keepership of both the Ashmolean Museum (from 1854) and the Oxford University Museum from its inception in 1857.

His death came at Oxford when, following dinner at All Souls College on the evening of 23 April 1874, he fell headlong down a flight of stairs, suffered a seizure and died the following day. His body was carried to York by train where it lay 'in state' in the Yorkshire Museum for a night, before being laid to rest in the local cemetery, appropriately beneath a thick slab of Coal Measures sandstone.

As early as 1834 Phillips was thinking about sediment and how it would be possible to determine its rate of deposition, and in 1839 in his book *Treatise on Geology*, he discussed the role rivers played in moulding the surface of the Earth through denudation. In 1858 Phillips was elected President of the Geological Society and he held this position until 1860. Like all presidents before and since, it was part of his duty to present medals to various notable and usually deserving geological savants at the Annual General Meeting. Accordingly on 17 February 1860 in front of the assembled Fellows, he took the chair to preside for the second time over the annual meeting. Searles Valentine Wood Snr (1798–1880), a former Ship's Officer with the East India Company, received the Wollaston Medal for his work on the fossils of the Crag (a horizon in the Pliocene), and in response he apologetically commented that its award should normally result in renewed enthusiasm for further work, but that on account of his age he thought that this was unlikely. In a rather sad aside he noted that he had been born in sight of a Crag pit (quarry), and would probably be buried within sight of another. From this one surmises that he thought he did not have much time left: in fact Wood was to live for another twenty years and the final part of his major work *The Crag Mollusca*, first begun in the 1840s, was not published by the Palaeontographical Society until 1882, two years after he had been lowered into the soil overlying his favourite geological horizon.

Following the civilities Phillips delivered his anniversary address and read out a list of recently deceased members that included the unlikely pairing of Alexander von Humboldt (1769–1859) and Archduke John of Austria (1782–1859), the youngest brother of Emperor Francis I. John Baptist Joseph Fabian Sebastian von Habsburg, to give him his full name, apparently promoted geological and mineralogical studies of the Austro-Hungarian Empire and was 'always happy to welcome the English visitor who carried a hammer and sketchbook.' He would not have afforded such a cordial welcome to French geologists in the year before his death: that year the two Empires clashed and the Austrian army partially under his command was first defeated at Montebello on 20 May and soon afterwards elsewhere. On 6 July 1859 Austria surrendered to the French.

Phillips then got to the core of his address, which was partly a review of recent geological work, but also an account of some of his own ideas. Of interest to us now are the sections headed 'Geological Time' and the immediately following 'Conversion of Geological into Historical Time'. In the first he stated that the thickness of the sedimentary pile as seen in the various strata amounted to 72,584 feet. The Palaeozoic was 57,154 feet thick, the Mesozoic 13,190 feet thick, and the Cenozoic 2,240 feet thick. These thicknesses he converted to the percentage of geological time that they represented: 79% Palaeozoic, 18% Mesozoic and 3% Cenozoic. He stated: 'It is possible by some hypothesis of the annual waste of the surface of land, or the annual deposition of sediment, as now observed in the sea, at the mouths of rivers or in lakes, to transform the unit of geological time above suggested into an equivalent term of years.' Tantalisingly he did not state what his figure was.

As an example he looked at the denudation of the rocks that formerly covered the rocks of the Weald of Sussex, an area of 3,000 square miles that includes Sussex, much of Kent and some of Surrey and Hampshire. Geologically the district is underlain with a variety of terrestrial sediments 1,000 feet thick of muds and sands, generally termed the 'Wealden', but including distinct horizons such as the Weald Clay, the Tunbridge Wells Sands and the Gault (Figure 10.2). Phillips could have taken any geographical region for his example but

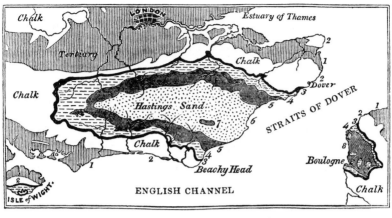

1. ▦ Tertiary.
2. ☐ Chalk and Upper Greensand.
3. ▬ Gault.
4. ▤ Lower Greensand.
5. ▩ Weald clay.
6. ▨ Hastings sands.
7. ▨ Purbeck beds.
8. ▨ Oolite.

Figure 10.2 Geology of the Weald, southeast England, showing the extension of the strata across the English Channel into France. Chalk (2) forms the hills of the North and South Downs while older softer rocks are exposed inbetween (from C. Lyell, *Manual of Elementary Geology* (1855), p. 273).

he took the Wealden district for a very particular purpose: his opposition to the conclusions on the age of this area announced a year earlier by Charles Darwin (Figure 10.3) in the first edition of *Origin of Species*.

In his book Darwin had set out the problem (in the text between pages 285 and 286): picturing himself standing on top of the chalk North Downs, Darwin could visualise the high anticlinal form of the rocks that had once overlain the Weald district between the escarpment on which he stood and the South Downs thirty miles further south. The area was now of lower elevation than the Downs, but was known since the days of John Farey to have once been an anticline of folded rocks that extended across the English Channel into the Bas Boulonnais district of France. Darwin wondered how much rock must have been removed to produce its present-day physiographical

Figure 10.3 Charles Darwin
(1809–1882). Photograph by
Elliott and Fry, 55 Baker Street.
Author's collection.

features. Andrew Crombie Ramsay (1814–1891), later Director of the Geological Survey of Great Britain, who was responsible for the figure of 72,000 feet for the British sedimentary pile, had told Darwin that several formations, or distinctive horizons, were on average 1,100 feet thick. Now he produced his startling conclusion: that as denudation rates were 1 inch per hundred years in his estimate, it was obvious that it had taken 306,662,400 years to produce the Wealden landscape, a process that had been ongoing since the Mesozoic. Three hundred million years! This was a fantastic figure, and yet Darwin said that the stripping off of this material was 'a mere trifle, in comparison with that which has removed masses of our Palæozoic strata', a time-span he did not attempt to quantify. Perhaps this was just as well, because it would have been far greater than the 300 million years for the Weald. Phillips, and many other geologists, could not fathom this conclusion at all. He could not accept the main Darwinian thrust, that of natural selection and evolution, and in reaching this conclusion he was not alone amongst geological colleagues. It suited Darwin nicely that the Earth was old: this meant that there was plenty of time available for natural selection and evolution to take place. For committed religious men of science this just could not be right. As a result of the adverse reaction that his chronological calculation provoked, Darwin reacted by reducing the significance of the passage in the second edition of *Origin of Species* and by the third edition, published in 1861, it had disappeared altogether.

Less than three months after his London address Phillips was in Cambridge where he delivered the annual Rede Lecture at the invitation of the Vice-Chancellor. This series had been established in 1524 using an endowment provided by the estate of Henry VIII's Lord Chief Justice Sir Robert Rede (d. 1519), and was considered to be one of the major events on the University calendar. Spurred into further thoughts following the London meeting Phillips took part of his 1860 presidential address and expanded it. Further augmentation of the text followed and Phillips published it as *Life on the Earth: its Origin and Succession*, a slim but important volume of 224 pages published by Macmillan of Cambridge and London. In a letter addressed to Charles Lyell dated 18 May, Darwin told him that the Reverend John Stephens Henslow (1796–1861), his great friend and teacher, had informed him of Phillips' Rede lecture and remarked that he (Phillips) had treated *Origin of Species* fairly. Of *Life on the Earth*, Darwin was not complimentary: responding in January 1861 to a letter that he had received from Joseph Dalton Hooker (1817–1911), a major ally and later Director of the Botanic Gardens at Kew, he sardonically agreed that Phillips' book was 'unreadably dull'.

In *Life on the Earth*, Phillips recognised that 'the Geological Scale of Time is founded on the series of strata deposited in the ancient sea.' He gave the results of his denudation calculations, and reported that the sedimentary pile that comprised the Cambrian and later sediments had begun to form anything between 38 and 96 million years ago. He argued that the denudation rate of 1 inch per century cited by Darwin was incorrect and that the rate would have been closer to 1 inch per annum. Taking the Ganges and Brahmaputra rivers and the Bay of Bengal as an example he said that the ocean was fed with 6,368,077,440 cubic feet of sediment each year which would cover the floor of the Bay of Bengal with just under one-hundredth of an inch. Taking Ramsay's thickness he converted it into inches and the resulting division of 864,000 inches by the annual thickness of sediment deposited by the Ganges/Brahmaputra gave 95,904,000 years. He did note some problems with this method and was aware that the monsoon effects on erosion dumped more sediment into the Ganges/ Brahmaputra than would be found in most other rivers. He then modified the length of time because the surface of the Earth had been 20 °C hotter in the past that it was at the present. This hotter environment would have created more atmospheric moisture and so the effect of rain and river flow would have been greater in the past. This reduced the time needed to deposit the Cambrian to Recent sediments by nearly 32 million years to 63,936,000 years. Finally, further modifications to this number were needed to take into account the differences in tectonic activity between the ancient past and younger times. Phillips argued that the surface of the Earth in its infancy was prone to greater uplift, movement and folding and therefore more rocks were subjected to erosional forces. This necessitated reducing the figure to 38 million years.

In other words, although the calculation was not quite so simplistic, Phillips had determined the thickness of the sedimentary pile and using various figures for the varying rates of sedimentation through time had arrived at this figure. In terms of scientific history this was an important calculation. It was the first time that such a method had been employed for the purpose of determining the Earth's age. It also came at a time when other scientists, equally displeased with Darwinian ideas, were devising thermal methods of dating the Earth. This methodology was championed by William Thomson, who was supported in his views by John Phillips.

In respect of the Weald both men were incorrect in dating the onset and duration of its denudation. It is now known that the area was first uplifted into a broad dome only twenty-five million years ago during a late episode of continental tectonic activity at the end of the Alpine mountain-building episode or orogeny. Phillips, quite by accident, was closer in his estimate than Darwin.

In *Life on the Earth*, Phillips' belief was set out early in the volume on page 3 thus: 'Nature, in a large sense, is the expression of a DIVINE IDEA, the harmonious whole of this world of matter and life. Man, included in this whole, is endowed with the sacred and wonderful power of standing in some degree apart.' As Jack Morell has noted in the *Oxford Dictionary of Biography*, Phillips 'reaffirmed his belief in divine design, the reality of species, the relative novelty of humans, and a reverential reading of the book of geological strata.' Convinced that Darwin had exaggerated the incompleteness of the fossil record, Phillips believed that its discontinuities were explicable only in terms of separate creations which were transcendental and inscrutable acts of God. He was 'unhappy with Lyell's advocacy of the antiquity of man.' Phillips warned his fellow geologists that investigating geochronology would be futile: 'Let him look at the Mosaic narrative, and be satisfied with the truth, that "In the Beginning God created the heavens and the Earth", for no measure of time conceivable by man will reach back to that remote epoch in the history of our solar system.'

WAS DARWIN A GEOLOGIST?

Darwin is well known for his biological and evolutionary ideas, but did he have any understanding of geology? Could he have been called a geologist? If he was, did Phillips consider Darwin to be less qualified than himself, and did this give him the moral high ground? Perhaps Phillips did, but interestingly neither man followed a university geological course to any great depth. In fact Phillips did not matriculate at all. In 1825 Darwin first attended the University of Edinburgh where as a diversion to his medical studies he attended the lectures in Edinburgh of Robert Jameson during the session 1826–1827 and found them 'incredibly dull. The sole effect they produced on me was the determination never as long as I lived to read a book on Geology, or in any way study the science.' Obviously at the time, the University of Edinburgh did not have a student assessment programme, which was fortunate for Jameson, or an Office of Teaching Methodology, which was unfortunate for his students. Fortunately for science, Darwin could not continue his medical studies as he was unable to cope with observing operations, which were conducted at the time without the aid of anaesthesia, and he dropped out in 1827. Returning home he was soon afterwards sent to Cambridge where he was encouraged to take a general degree – the usual route for those seeking a career in the Church – and while there came under the influence of Henslow, the University Professor of Botany, and Adam Sedgwick, the Professor of Geology. Thanks to these two men, Darwin developed a serious interest in natural history. In July 1831 he had purchased a clinometer used to measure the dip and strike of layers or beds of rock and at much the same time attempted to draw a crude geological map of the district around Shrewsbury. The following month he spent some time in the company of Sedgwick rambling through the geology of the north Wales coastline. He then joined the *Beagle* as the ship's naturalist. In his library on board was Lyell's *Principles of Geology* which he read and was convinced by Lyell's premise of the uniform state of geological processes. He carried out geological field work on the Azores, and collected fabulous fossils in Patagonia. He wrote up his geological observations made on the voyage in three works: *The Structure and Distribution of Coral Reefs* (1842); *Geological Observations on the Volcanic Islands, Visited During the Voyage of H. M. S. Beagle* (1844); and *Geological Observations on South America* (1846).

Thanks to his writings and observations made during the *Beagle* voyage, Darwin's scientific reputation was without question. Although the majority of people today regard him as a biologist, the bulk of his early works were in fact geological. He was regarded as being a geologist by his peers of the Geological Society, which in 1859 awarded him the Wollaston Medal. This was presented at a meeting on 18 February 1859 and was handed over by none other than the President for that year, John Phillips (who himself had been its recipient in 1845). In his address, Phillips remarked that Darwin had 'never ceased to labour ... in the cause of geology', and 'through great tracts of America his masterly hands have sketched and measured the prominent structures of rocks' and that his work 'has added much to a reputation already raised to the highest rank.' Unfortunately on account of poor health, Darwin could not be present to receive the medal in person and Charles Lyell was given the medal on his behalf. Through his great friend, the recipient stated that the medal was 'more prized by him as a mark of your sympathy, because it cheers him in the seclusion in which he finds it necessary to pursue his studies and researches'.

PROBLEMS WITH PHILLIPS' IDEA

If one returns to Phillips' novel dating methodology one finds that there was a major difficulty with it: how could one determine accurately the actual thickness of sediment deposited? Phillips stated that 72,584 feet of sediment had been deposited since the beginning of the Cambrian. Of the twenty or so estimates of sediment thickness that were used to estimate the age of the Earth between 1860 and 1927, the figure of 12,000 feet of sediment used by James Croll (1821–1890) in a paper published the year before his death was the lowest, while the highest was 335,800 feet given by William Johnston Sollas (1849–1936) in 1909, although like some other authors he included the significant thickness of Precambrian deposits in his calculations (see Table 10.1). However, Arthur Holmes in 1927 in his important book *The Age of the Earth: An Introduction to Geological Ideas*, number 102 of Benn's Sixpenny Library, a series that its publishers proclaimed 'has the revolutionary aim of providing a reference library to the best modern thought, written by the foremost authorities', preached caution. He suggested that the total sediment pile was closer to 529,000 feet in thickness, and noted that 'most of these [earlier] estimates are little more than rough guesses. We do not know how much of the story is lost to us, or how much is hidden away.' Holmes

Table 10.1 *Various estimates of the age of the Earth derived by the sediment accumulation method.*

Date	Authority	Maximum sediment thickness [^aincluding Precambrian] (in feet)	Rate in years for depositing [^bor eroding] one foot of sediment	Age of the Earth [^cor time since start of Cambrian] (millions of years)
1860	J. Phillips	72,000	1,332	54 (38–96)
1869	T. H. Huxley	100,000	1,000	100
1871	S. Haughton	177,200	8616	1,526
1878	S. Haughton	177,200	–	200
1880	A. R. Wallace	177,200	158	28
1883	A. Winchell	–	–	3
1889	J. Croll	12,000	6,000[b]	72
1890	A. De Lapparent	150,000	600	90
1892	A. R. Wallace	177,200	158	28
1892	A. Geikie	100,000	730–6,800	73–680
1893	W. J. McGee	264,000	6,000	1,584
1893	W. Upham	264,000	316	100
1893	C. D. Walcott	–	–	45–70
1893	T. M. Reade	31,680	3,000[b]	95[c]
1895	W. J. Sollas	164,000	100	17
1897	J. G. Goodchild	–	–	704[c]
1897	J. J. Sederholm	–	–	35–40
1899	A. Geikie	–	–	100
1900	W. J. Sollas	265,000	100	26.5
1909	W. J. Sollas	335,800[a]	100	80
1909	J. Joly	265,000	300	80
1914	J. Joly	–	–	87

took the figures given by earlier workers and came up with eight different results that ranged from 80 million to 350 million years, a variation of 270 million. Rightly, he recognised that parts of the Phanerozoic sedimentary record might have been eroded away, but precisely how much sediment had been lost was difficult to determine. If an unconformity is recognised in the field, such as those seen by James Hutton in Scotland, it is not always obvious what this sedimentary interregnum represents in terms of time and lithology.

LATER ESTIMATES OF TIME DERIVED FROM
THE SEDIMENTARY PILE

Further hour-glass calculations using Phillips' method followed. In 1868 Archibald Geikie (1835–1924), the then Director of the Geological Survey of Scotland, published an account of present-day denudation rates. He pointed out that given the rate of reduction of the present land surface by 1 foot in every 6,000 years, Europe would disappear in four million years, North America in four and a half million years, and both Asia and Africa in seven million years. He recognised that physicists had expressed some difficulties with the age of the Earth as it had been calculated by their scientific colleagues the geologists, and warned that the estimates made by the latter might have to be reduced, contrary to geological evidence. Perhaps the most celebrated work in this area (on account of the huge fluctuations of his time determinations) was that by the Dublin geological professor the Reverend Samuel Haughton (1821–1897) who, like John Phillips, was a supporter of William Thomson and an opponent of Charles Darwin. Haughton had interests in many fields: geology, mathematics, animal physiology, medicine and education. Born to Quaker parents, he was brought up an Anglican and educated at Trinity College Dublin, where he became a Fellow at the remarkably tender age of twenty-four. Almost immediately he was ordained, and was appointed to the Chair of Geology in 1851, the chair filled by Phillips eight years earlier. During his tenure he also studied medicine at the College during which time he developed his keen interest in animal physiology.

It was said that it was not unknown for him to grab his case of surgical instruments in the middle of a lecture and head towards the Zoological Gardens on hearing of the death of a hippopotamus or another large mammal. He is fondly remembered by some for having devised a humane method of hanging: he calculated the length of drop required to effect the instantaneous death of the condemned. Prior to this many criminals simply suffered slow strangulation.

In 1865, Haughton in the first edition of his *Manual of Geology* wrote that the Earth was 2,298 million years old, a calculation that he based on the same global cooling principles of those of William Thomson (whose exploits are discussed in the next chapter). In the context of the sediment accumulation method of age determination, Haughton is remembered for his 1878 principle that 'the proper relative measure of geological periods is the maximum thickness of the strata formed during these periods': this of course necessitated a global estimate of sedimentary sequences. In 1871 in the third edition of his *Manual of Geology*, he published a date of 1,526 million years, based on denudation rates; and this date was revised by him seven years later to 200 million years in a paper published in *Nature* in which he tried to prove that past climatic changes were not due to alterations in the position of the Poles – still a topical subject today. Much later in 1947 Arthur Holmes argued that the uniformity of sedimentation rates assumed by Haughton was incorrect and that his principle would be better stated as: 'the time elapsed since the end of any geological period is a function of the sum of the maximum thicknesses accumulated during all the subsequent periods.'

In the 1890s, the methodology was revisited by several authors in the United States, including Charles Doolittle Walcott (1850–1927). In 1893 he presented a paper at a meeting of the American Association for the Advancement of Science at Madison, Wisconsin, in which he gave a timespan of 60 to 70 million years for post-Archean time (the Archean was the time prior to the first appearance of shelly fossils). His work was based on accurately measured sections in North American sedimentary basins (Figure 10.4). He was fortunate, as

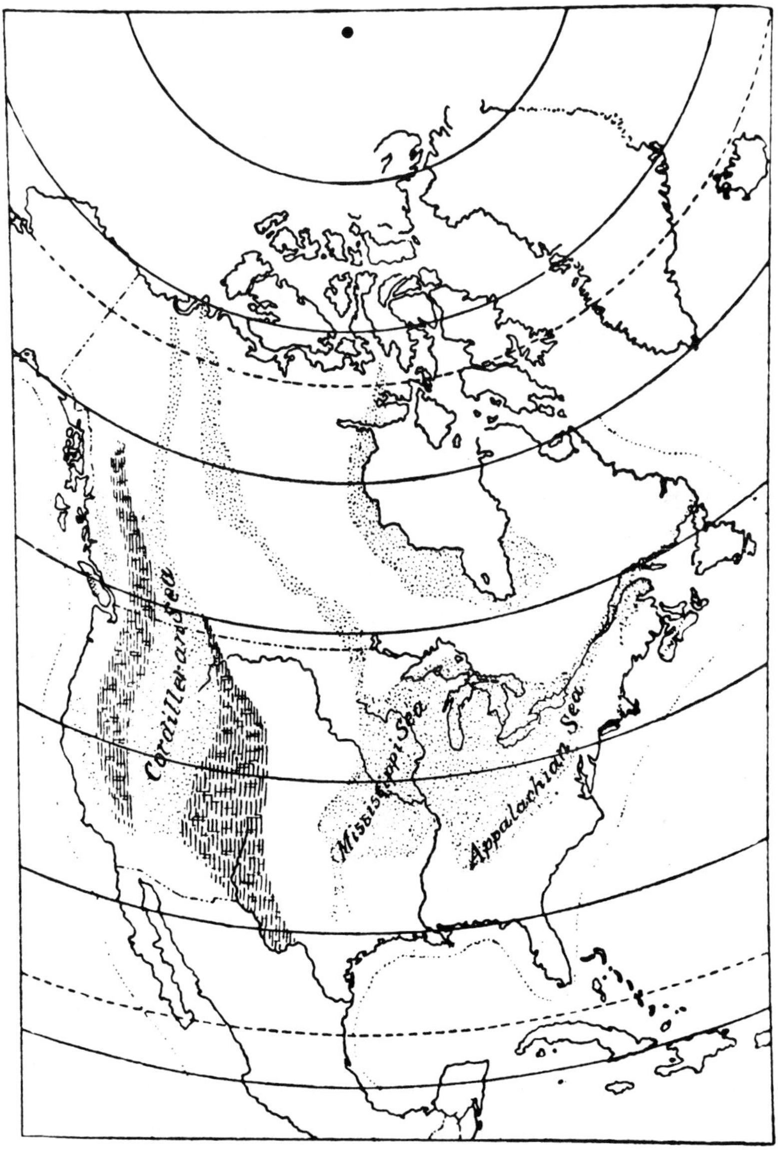

Figure 10.4   Charles Doolittle Walcott's 1893 map showing the positions of former seas that covered the North American continent. Measuring thicknesses of limestone found in the most western Cordilleran Sea and determining their annual rate of deposition gave Walcott an estimate of the age of the Earth.

Ellis Yochelson his biographer has pointed out, that the rocks of the American west were generally better exposed and easier to measure than those in Britain. He was also responsible for the word 'geochrone' from which the term 'geochronology' evolved. Walcott's paper appeared in the first volume of the recently established periodical *Journal of Geology*, and it achieved a great and wide circulation thanks to being republished shortly afterwards in the *American Geologist*, the *Proceedings of the American Association for the Advancement of Science*, and the *Annual Report of the Smithsonian Institution for 1893*.

Walcott had a brilliant career, rising from the lowly position of farm labourer to become the Director of the United States Geological Survey in 1894 and then from 1907 Secretary of the Smithsonian Institution in Washington, which was probably the greatest scientific research facility of the time. He is now best remembered for his startling discovery in September 1909 near Mount Wapta in British Columbia, Canada, of the celebrated Middle Cambrian lagerstätten, the Burgess Shale. Enthralled by the fossils that he found, Walcott returned for many field seasons and brought his wife and children along to assist in their excavation. All told he recovered 65,000 specimens that now are in storage in Washington where they continue to be studied. This remarkable fauna contains 170 species, including many, many examples of soft-bodied organisms the likes of which had never been seen before Walcott's work. This fauna and others discovered in Greenland and China more recently have given palaeontologists great insights into some of the earliest inhabitants of the world's Cambrian oceans.

Also writing in 1893, Warren Upham (1850–1934), a United States Geological Survey Quaternary geologist who specialised in studies on Glacial Lake Agassiz, suggested that given the great diversity of life a great deal of time was needed for their development, and so the stratified rocks represented a time-span of approximately 100 million years. A year earlier another American William John McGee (1853–1912), later noted for his anthropological and conservation work (who was known as 'no stop' McGee to friends and 'full stop' McGee to detractors because he styled himself 'W J McGee'), announced to a meeting of the American Association for the Advancement of Science that he estimated, using erosion as a measure, that the Earth was 15,000 million years old and that 7,000 million years had passed since the beginning of the Cambrian. The latter figure was remarkably large and quite at variance with others propounded at the time. The following year he admitted to some mistakes in his calculations and reduced the figures to 6,000 million years and 2,400 million years respectively.

Meanwhile across the Atlantic the Liverpudlian architect, civil engineer and part-time but talented geologist Thomas Mellard Reade (1832–1909), who incidently was a three-time President of the Liverpool Geological Society, took up the challenge of Phillips' method. He had, in his 1877 presidential address to the Society and in an expanded book version published two years later, discussed the denudation of soluble geological materials such as limestone, but later expanded his interests to examine the non-soluble sediments as well. He was to produce a series of papers in the *Geological Magazine* and elsewhere, of which his short contribution entitled 'Measurement of Geological Time' published just prior to Walcott's in 1893 stated that the Cambrian began 95,040,000 years ago. This paper was also important for his observation that it was difficult to calculate just what a sedimentary thickness represented chronologically: 'it may be reasonably objected that 10 feet of one set of strata may chronologically represent 1000 feet or more of another.'

Alfred Russel Wallace, in his important book *Island Life: Or, the Phenomena and Causes of Insular Faunas and Floras, Including a Revision and Attempted Solution of the Problem of Geological Climates* (Macmillan, 1880), wrote that 200 million years was all that was required for the development of the world's faunas. In its pages he also entered the sedimentological debate and took Haughton's cumulative thickness of the stratified rocks of 177,200 feet and remarked that they would have been deposited over a period of 28 million years. He repeated this in a paper published that year in *Nature*, a journal he had helped establish in 1869. This conclusion led to a flurry of papers in the *Geological Magazine* in which convoluted textural explanation, mathematical calculations, thickness estimates and sniping by all parties prevailed. Mellard Reade was the first to enter the fray in 1883 remarking that 'it is evident that the origin of Mr Wallace's difficulties is that he does not realize fully the conditions of the problem he set himself to solve.' Reade was followed soon afterwards by a Sydney B. J. Skertchly, and nine years later Bernard Hobson dared to revisit Wallace's geochronological estimate and outlined his misgivings in the pages of *Nature*. Wallace stood his ground: 'any such general examinations of this question from an adverse point of view, I have hitherto failed to meet with.'

In 1895, Sollas, using Haughton's principle, calculated the Earth's age to be 17 million years, while two years later John George Goodchild (1844–1906), a geologist with the Geological Survey of Great Britain, estimated that 704 million years had elapsed since the beginning of the Cambrian. Goodchild's figure is considerably higher when compared with those (excepting that of McGee) published by others around this time. Sollas returned to the problem in 1909, when his calculations based on a total sediment thickness of 335,000 feet or 63 miles and sedimentation rates of 3 and 4 inches per century gave him an age of the Earth of 148 or 103 million years. However, he noted that it was difficult to determine accurately the rate of sediment accumulation, which he acknowledged could be anything between 2 and 12 inches per century. He returned to the problem in 1900 and again in 1909 when as President he addressed the Geological Society of London.

John Joly (1857–1933), who had succeeded Sollas as Professor of Geology and Mineralogy at Trinity College Dublin, was a leading player in the geochronological debate largely on account of his 1899 sodium method paper (this will be discussed in Chapter 12). He entered the sediment accumulation debate in 1909 when he examined Sollas' figures for himself, and in his subsequent publication of 1911, wrote that he agreed with his predecessor's result. However, in 1914, in a lecture to the Royal Dublin Society he argued that sediment mass, not thickness, was a more accurate measure, and this yielded a minimum of 47 million years and a maximum of 188 million years. He reduced the mean of these limits to a figure of 87 million years on the basis that he believed sedimentation rates were not uniform through geological time.

THE HOUR-GLASS SHATTERS

It is clear that the nature of sediment production and its subsequent deposition is dynamic and controlled by an infinite number of factors that are difficult to quantify. Using sedimentation rates and accumulation as a geological chronometer, geologists and biologists did come up with dates for either the beginning of the Cambrian or the age of the Earth, and reached a general consensus by 1900 that 100 million years or less for this method was a reliable estimate. Although one could scoff at the apparent foolishness of those who tried this calculation, such as its instigator John Phillips and his later disciples, one cannot deny that the sediment accumulation concept which he first considered in the mid 1830s was an interesting one, and no more absurd to its advocators than other measures used earlier.

By 1910 these sedimentary chronologies were being supplanted by the findings and age determinations generated by radioactive decay methods. Although in its infancy, the study of radioactivity was beginning to yield Earth ages that were considerably older than the 100 million years suggested by the sediment accumulation measure. Even sedimentological chronologies were being revised upwards. At a meeting of the Geological Society of America held in Albany, New York, in 1916 the Yale professor Joseph Barrell (1869–1919) presented a paper on 'Rhythms and the measurement of Geologic Time' that was later published in full in the Society's *Bulletin* in 1917: it has since achieved classic, almost cult, status amongst geologists. This may be partly due to the fact that its author died young only three years later, but also due to its revolutionary content in which he suggested that the new radiometric dates should be used to interpret the sedimento-logical record. Barrell examined in detail the various methods used to date the Earth to that time, and their resulting figures, and in Part 4 of the paper concluded that a minimum of 550 million years and a maximum of 700 million years must have elapsed since the beginning of the Cambrian, and that the underlying igneous and sedimentary rocks termed the Laurentian were at least 1,400 million years old. The Earth, he surmised, was much older even than this.

11 Thermodynamics and the cooling Earth revisited

In the middle of the nineteenth century numerous oceanographic voyages traversed the world's oceans revealing details of submarine topography and hitherto unseen animals. At much the same time telegraphic communications were being developed and the network in Britain expanded rapidly in the 1830s as the railways were laid out. By 1850 Britain had telegraphic links with France and Ireland, but it was soon realised that theoretically it could be possible to link Europe with distant continents including North America and Africa. In 1858 the first transatlantic cable was laid between Europe and North America but a month later problems with the insulation led to signal failure. In 1865 a new attempt to lay a cable followed but the line was lost. Undaunted, the steamer the *Great Eastern* set out on 13 July 1866 and began to lay a working cable for 2,000 miles between Valentia Island off southwest Ireland to Heart's Content, Newfoundland. Unusual for the time, this ship was powered by both paddles and screw propellers. It completed its voyage and the link by 27 July and a message was sent from Canada to Edward, Lord Stanley, the Prime Minister. The following day Queen Victoria sent a message in the opposite direction to Andrew Johnson, the President of the United States, from Osborne House and expressed her hope that the cable might 'serve as an additional bond of union between the United States and England'. Much of the credit for the success of telegraphy is owed to a Belfast man, William Thomson (1824–1907) (Figure 11.1) who was a director of the Atlantic Telegraph Company, and rewarded for this work with a knighthood in 1874. Thomson carried out a great deal of research on cables, determining the diameter required and the purity of copper necessary to ensure that they did not malfunction, and he also invented a submarine telegraph receiver which allowed the incoming message, in Morse Code, to be recorded.

Figure 11.1 William Thomson, later Lord Kelvin (1824–1907) in 1854 (from his obituary in *Proceedings of the Royal Society of London*, Series A, **79** (1907), iii–lxxvi).

He is probably better known to most as Lord Kelvin of Largs following his elevation to Baron in 1892. The name Kelvin came from a small river that rises in the Kilsyth Hills, flows for 21 miles close to Glasgow and empties into the River Clyde near Partick; and Largs from the holiday resort in Ayrshire where he built his seaside mansion Netherhall using earnings gained from his work connected with the laying of the transatlantic cable. He had designed much of the house himself and had it fitted with electric light. Additionally he was able to buy an 82-foot-long, 121-ton schooner, the *Lalla Rookh*, in 1870. He used this vessel primarily for recreational sailing but also carried out research on methods of depth sound, and recommended the use of piano wire for recording depths. This method was used during the celebrated oceanographic research voyages of the HMS *Challenger* in 1872 but was soon found to be difficult to handle and so was abandoned shortly into the voyage.

William Thomson was born in Belfast on 26 June 1824, son of a professor of mathematics. It soon became obvious that he was very intelligent and his ambitious father ensured that both he and his older brother James (1822–1892) studied hard. Perhaps this was in response to the loss of their mother when William was just six years old. James in due course became Professor of Civil Engineering at the University of Glasgow, a Fellow of the Royal Society and inventor of the vortex turbine. In 1832 the family moved to Glasgow where William was to spend much of the remainder of his life. He was enrolled at the university at the tender age of ten, but did not graduate. Instead he furthered his studies in Cambridge from 1841 where he enhanced his reputation as an athlete and mathematician, and followed this period with further study in France where he became interested in heat. This was a research interest that was to remain with him throughout his career. In 1845 he was appointed Professor of Natural Philosophy (what we would now call physics) at his alma mater and so for a period was on the faculty with his father who had canvassed hard for his son's appointment.

Thomson was somewhat unlucky in his personal life. When he decided that he wished to marry he proposed to Sabrina Smith three times, and three times she rejected his advances. Undaunted, he turned his attention to the daughter of a family friend: three months after his third rejection by Miss Smith he proposed to Margaret Crum, and soon afterwards they married. Almost immediately she succumbed to serious illness and remained an invalid, bed- or sofa-bound until her death in 1870. Four years later at the age of fifty he married Frances Anna Blandy whom he had met in Madeira while engaged in some oceanographic research. Thomson had no children – perhaps this is why he could devote so much time to his scientific investigations.

On his return to Glasgow in 1845 he established a laboratory in which students were trained in scientific methods and techniques. He believed that they would become usefully employed in industrial enterprises that were springing up in the region and making Glasgow a major industrial centre. He then embarked on research on heat, being interested in the Second Law of Thermodynamics which had been formulated in 1850 by Rudolph Julius Emanuel Clausius (1822–1888). Thermodynamics is defined in some dictionaries as 'the science of heat as a mechanical agent', and the second law is concerned with the directional flow of heat. Thomson's ideas on heat and its ability to be transformed into directional force were to have an important bearing on his geological and geochronological ideas with the passage of time.

THOMSON ON THE AGE OF THE EARTH: A THREE-PRONGED ATTACK

Infuriated by Charles Darwin's dabbling in geochronology, Thomson set out to prove through the application of the laws of physics the *actual* age of the Earth, and even sought the views of John Phillips as to the validity of Darwin's geological conjectures. We have a good idea as to what Phillips's reaction would have been. The chronological issue of the Earth was tackled by Thomson in three research strands: the first was in relation to the Sun – he attempted to estimate how long it had been shining and used this as a corollary for the age of the Earth; the second took the secular cooling rate of the Earth, thus revisiting Comte de Buffon's work nearly a century earlier; the third involved an investigation into the effect that friction caused by tides might have had on the shape of the Earth. Three very different schemes, but they ultimately led to much the same conclusion in terms of age determination.

The age of the Sun

Thomson produced a large volume of work on the Sun and its heat, and first published on the subject in 1854. He suggested that as the Sun formed thanks to the collision of meteorites which built up its mass, the gravitational energy that pulled them towards the Sun was released as heat. The output of heat, he argued, was far greater during the early life of the Sun than it was at present, and in an aside pointed out that in his opinion 2 million years was an adequate length of time for the uniformitarian geological processes that had been described by geologists such as Charles Lyell. Eight years later he argued that the Sun 'has not illuminated the earth for 100,000,000 years, and almost certain that he has not done so for 500,000,000 years', and that it was probably closer to the truth that the Sun had been operational for between 20 and 60 million years. Towards the end of his life, following persuasion by his associate Peter Guthrie Tait (1831–1901) he argued that the Sun was no older than 20 million years. This by association gave a limit to the age of the Earth.

Cooling of the Earth

Like Buffon, Thomson believed that the primordial Earth was molten throughout, and that it solidified from its centre outwards as the internal heat migrated through the rocks by conduction and heat was lost from its surface. Once solid, but still hot, it lost heat by conduction rather than convection, which sees heat carried through a fluid volume by a flow of matter. In 1862 he carried out experiments to determine the conductivity of various rock types, and took 7,000 degrees Fahrenheit as the temperature of fusion of rocks, an estimate that had been determined a short time earlier. In this research he was greatly influenced by the work of French physicist Jean Baptiste Joseph Fourier (1768–1830), which he had studied during his continental sojourn in 1845. Fourier had written that the source of the Earth's heat was three-fold: from primitive internal heat, from heating by the Sun, and from heat in the Universe. Thomson also drew on work that had revealed the temperature gradient in the Earth and noted that temperatures increased approximately one-fiftieth of a degree Fahrenheit per foot that one went down. This had been investigated by taking temperature readings from various depths in mines or from boreholes. This work was not very accurate until the invention in the 1830s of specialised thermometers designed for the task. Even by the 1860s results were not very conclusive, and it was believed that the temperature at the Earth's centre was approximately 3,800 degrees

Centigrade. We now consider the temperature at the centre of the Earth to be about 7,200 degrees Centigrade. In 1862 Thomson stated in the *Transactions of the Royal Society of Edinburgh* that the Earth was somewhere between 20 and 400 million years old, with 98 million years being the likely age. He argued, given the underground temperatures known and the loss of heat due to conduction, that if the Earth was as much as 20,000 million to 30,000 million years old the underground temperatures observed should have been far lower. It was possible however to generate the observed temperatures if the whole of the Earth's surface had been heated up to 100 degrees Centigrade at some time in the past 20,000 years. Naturally this could not have happened as all traces of life would have been killed.

Six years later, in an address to the Geological Society of Glasgow delivered on 27 February 1868, he concluded that the Earth was no more than 100 million years old, and this timescale was questioned the following year by none other than Thomas Henry Huxley (1825–1895) who asked in his presidential address to the Geological Society, 'Has it ever been denied that this period *may* be enough for the purposes of geology?' In doing so Huxley laid the seeds of discontent between the views of physicists in one corner and the geologists and biologists in the other corner, a disagreement that continued for nearly half a century. In 1895 Thomson penned a short paper for *Nature* in which he discussed a recently published paper by Clarence King (1842–1901), who had been the first director of the United States Geological Survey between 1879 and 1881. King's paper re-examined Thomson's work and using a new figure of 1,950 degrees Centigrade for the fusion temperature of rocks, a figure which he had obtained from the physicist Carl Barus (1856–1935), arrived at 24 million years as the age of the Earth. This limit was accepted by Thomson who acknowledged that Barus' data would have rendered his 100-million-year limit too large, and that while the method would point to 10 million years, the effects of pressure on the geological processes would push the age determination to that of King. In a short letter to *Nature* in 1897, which gave his final pronouncement on the cooling Earth method, Thomson settled on a range of 20 to 40 million years old, but said that the 24-million-year estimate reached by King in 1893 was probably correct.

Tidal friction

Thomson's final methodology for fixing the age of the Earth appeared in 1868 in a paper published by the Geological Society of Glasgow, and it relied on changes in the Earth's shape over time, changes effected by a lowering of its rate of rotation caused by tidal friction. It was well known that because of friction on the surface waters as the Earth spins, tidal waters tend to become banked up and do not act in a predictable manner normally expected. The Earth, he said, assumed its flattened spherical shape soon after its formation while it was still molten. He realised that if one took the present rotation rate of the Earth, and used this to calculate what the shape of the globe would have been if this had been the primordial spinning rate, one would expect a spheroidially flattened globe of a particular shape. This expected shape he found was not appreciably different from the actual shape of the globe, and so he deduced that very little time had elapsed since the formation of our planet. He was however unprepared to give an actual time limit based on this formulation.

REACTION TO SOLAR HEAT, A COOLING GLOBE AND A SPINNING SPHERE

Thomson's final contribution to the chronological debate was published in America in 1898 in a paper entitled 'The age of the earth as an abode fitted for life'. This was reprinted in several journals in England in 1899 and is basically a summation of his ideas. He held on to his view that the Earth had consolidated no more than 40 million years previously, and that the Sun was no more than 20 to 25 million years old. His invocation of a solid cooling Earth was at variance with the views of other scientists, but not all. Samuel Haughton, who supported his fellow Irishman's work, applied the cooling method in 1865 and in his *Manual of Geology* published his calculations in which he arrived at an estimate of 2,298 million years for the age of the Earth. Haughton was no mathematical slouch and his calculations were spot-on, but his problem was that the basic figures he used for the calculation were flawed.

By 1868, following work by Charles Darwin's fifth child George Howard Darwin (1845–1912) on the orbit of the Moon and the Earth's retardation rates, Thomson stuck his head above the parapet regarding the tidal friction estimate of time and said that on the basis of this method the Earth was less than 1,000 million years old. Darwin *fils* had suggested that the Moon had been formed by metastasis from the Earth, from material thrown out owing to rapid spinning of the parent body – a theory now known to be fantastic and incorrect. He calculated the time taken for the Earth and Moon to settle down from this initial rupture to their present condition and came up with a minimum of 56 million years.

Thomson's detractors, of whom there were few, given his reputation, questioned his figures regarding the internal temperature and gradient in the Earth, and also questioned his reliance on a theory without what they considered any geological foundation. Thomas Mellard Reade in 1878 said that 'Facts are safer than theories' implying that Thomson relied too heavily on the latter. The Reverend Osmund Fisher (1817–1914) stated in a paper published in the *Geological Magazine* in 1895 that 'no reliable estimate of the age of the world, based on considerations of the present temperature gradient at the surface, has hitherto been made.' James Geikie (1839–1915), younger brother of Archibald and his successor to the Chair of Geology at Edinburgh, wrote in the February 1900 issue of the *Scottish Geographical Magazine*, 'there are certain other considerations which increase one's doubts as to the adequacy of Lord Kelvin's theory.' Thomas Chrowder Chamberlin (1843–1928), President of the University of Wisconsin, was probably Thomson's most outspoken critic, at least in America. He believed that the Earth had formed thanks to the accretion of cold material and that it had never been fully molten. Caustically he remarked in 1899 in a paper published in the journal *Science* that 'the postulate of a white-hot liquid earth does not rest on any *conclusive* geological evidence'. Even the biologists were concerned that Thomson's age limits were too short for biological evolution. Edward Bagnall Poulton (1856–1943), the husband of Emily Palmer of the biscuit family, and also Hope Professor of Zoology at Cambridge, weighed in at the annual British Association meeting in Liverpool in 1896. As President of the Biological Section he could create quite a stir, and in his address he attacked the findings both of Thomson and those of some geologists.

The most serious objection to some of Thomson's conclusions came from another Úlster Protestant, John Perry (1850–1920), who coincidently spent 1874 working as a research assistant in Thomson's laboratory in Glasgow before moving to Japan. Following a stint as a professor at the Imperial College of Engineering in Tokyo, Perry returned to Britain. He took a position at the Finsbury Technical College in London and in 1896 moved to the larger Royal College of Science in the city. In 1894 Robert Arthur Talbot Gascoyne-Cecil, third Marquess of Salisbury (1830–1903), was President of the British Association for the Advancement of Science meeting held that year in Oxford. Although not a scientist, he was a Fellow of the Royal Society, and he was influential: people listened to what he had to say. He had recently twice served as Prime Minister, when his Conservative party held power alternately with William Gladstone's Liberal Party, and he would hold this office again for a further eight years between 1895 and 1902. In his address Salisbury attacked the basis of Darwin's theory of natural selection saying that there was not enough time for natural selection to have taken place. To back his assertion Salisbury depended on the age estimates and reputation of Lord Kelvin. Having read the printed paper, Perry wrote to Kelvin three times and outlined a number of objections to his work and gave some suggestions as to how the calculations could be improved, but he received no response. Not content to be brushed off, Perry sought and received support for his objections from other scientists and then felt forced to commit himself to print. The following year his objections were outlined in two papers in *Nature*. These were immediately countered by Kelvin in a note in the same journal, and this was followed by a third missive from Perry. Broadly speaking the younger combatant said that Kelvin's reasoning for a young Earth must have been flawed. As Brian Shipley has recently pointed out Perry showed that 'the faster heat was conducted outwards from the Earth's core, the *longer* it would take to obtain the present observed temperature gradient at the surface.' This was because Kelvin had based his calculations on the conductive properties of surface crustal rocks and had not included the denser sub-crustal rocks known to exist. *Longer* was counter to Kelvin's conclusion.

However, there was some support at this time, and it came from the American George Ferdinand Becker who in *Science* in February 1908 re-examined Thomson's methods and concluded: 'Notwithstanding the inadequacy of the data, I can not but believe that the 60-million year earth here discussed is a fair approximation to the truth and that with better data this age will not be changed by more than perhaps 5 million years.' Becker pointed out that this age was broadly concurrent with those derived from the sedimentation accumulation and oceanic sodium methods. Even Mark Twain (1835–1910) who wrote a short essay on geological matters said of Thomson that 'I think we must yield to him and accept his view'.

ACCOLADES AND LAURELS

Thomson received many honours during his lifetime including at least seventeen honorary degrees, a knighthood, the Order of Merit (an honour established in 1902, and restricted to twenty-four individuals and the monarch), membership of the Order of Sacred Treasure of Japan, and the position of Honorary Colonel of the Electrical Engineers. The accolades culminated in his elevation to Baron in 1892, a title that became extinct on his death. Following this elevation, like all members of the aristocratic establishment, he had to adopt a coat of arms (Figure 11.2). The motto reads 'Honesty without fear', while the shield is supported on its dextral side by a capped Glasgow student resplendent in a scarlet gown and holding a marine voltmeter, and on its sinistral side by a sailor holding a sounding line and weight. It summed up Kelvin's attitude and interests in life. He

Figure 11.2 Kelvin's coat of arms adopted in 1892 (from Debrett's *Peerage* (1908)).

said and published whatever he liked without fear of what critical response it might engender, and he loved the contrast between the academic and maritime sides of his activities.

Following his death on 17 December 1907 Baron Kelvin, as he was by then, was buried close to other scientific luminaries in Westminster Abbey in London. He was commemorated in Glasgow, his adopted city, with a statue and in Belfast with another. Later still he was additionally honoured in the city of his birth with the erection of a blue plaque on the site of his birthplace. Unfortunately when ordering this plaque the Ulster History Circle fell into the trap that besets many a reader of the adventures of Tintin: how can you tell the bumbling detectives Thompson and Thomson apart? Sadly, they picked the wrong detective and the plaque reads 'William Thompson'; 'Thomson' is the Scottish variant of the name, while 'Thompson' is the English version. His peerage name 'Lord Kelvin' adorned the bow of a marine research vessel and the glossy cover of the novel *Lord Kelvin's Machine* by James P. Blaylock. However wonderful all these accolades may be, it is appropriate given his major studies on heat that he be remembered across the globe for the temperature scale that bears his name.

CONSIGNING THOMSON'S TEMPORAL TRIDENT TO HISTORY

Thomson's initial reasoning behind taking up research on the age of the Earth was his difficulty with the Darwinian timescale, and Darwin

himself acknowledged that it caused him trouble. If we believe
Archibald Geikie, his influence on the biologists who adhered to the
longer time frame required by natural selection would appear to have
been slight. In fact many biologists did not require the longer timescale
suggested by Darwin, as they did not accept natural selection. They
preferred the Lamarckian means of biological change which required
little time. Natural selection and thus the need for greater time was
generally accepted by most biologists by the 1930s. In his presidential
address to the geological community at the British Association for the
Advancement of Science meeting held in Dover in 1899, Geikie firstly
acknowledged the debt geologists owed Thomson:

> Geologists have been led by his criticisms to revise their
> chronology. They gratefully acknowledge that to him they owe
> the introduction of important new lines of investigation, which
> link the solutions of the problems of geology with those of physics

but then he delivered a sharp reprimand to the ennobled physicist and
a reminder of the leanings of palaeontologists towards a chronology
longer than the one that he proposed:

> It is difficult satisfactorily to carry on a discussion in which your
> opponent entirely ignores your arguments, while you have given the
> fullest attention to his. [Geologists] have been willing to accept Lord
> Kelvin's original estimate of 100 millions of years as the period within
> which the history of life upon the planet must be comprised ... yet
> there is undoubtedly a prevalent misgiving, whether in thus seeking
> to reconcile their requirements with the demands of the physicist
> they are not tying themselves down within the limits of time which
> on any theory of evolution would have been insufficient for the
> development of the animal and vegetable kingdoms.

Thanks to his continual tinkering with his age determinations
Thomson more than likely reduced the credibility of his work, and
his research in the area of geochronology was, like that of John Joly
and his oceanic salination scheme, overtaken by the emergence of

the study of radioactivity in geology. He failed to acknowledge fully this new science and to see its implications. It is perhaps too easy to castigate him for this and regard him as a scientific failure, and at the same time forget the scientific achievements that resulted from his brilliant intellect. Thomson was without doubt one of the pre-eminent scientists working in nineteenth-century Britain.

Shortly before Kelvin died, Ernest Rutherford (1871–1937), a young New Zealander who was Professor of Physics at McGill University, Montreal, was waiting to give a lecture on his work to an assembled audience at the Royal Institution in London in 1904. Looking out from the lectern Rutherford was horrified to see the 79-year-old Kelvin sitting in the front row, and he realised that the final section of his lecture would discredit a great deal of the older man's geological research on the age of the Earth. He had no option but to begin, and was highly relieved shortly afterwards to see Kelvin begin to slumber. Throughout the lecture Kelvin remained lost to the world but awoke at the critical point when Rutherford was about to deal with geochronology. He recalled later: 'I saw the old bird sit up, open an eye and cock a baleful glance at me! Then a sudden inspiration came to me, and I said Lord Kelvin had limited the age of the earth, *provided no new source of heat was discovered*. That prophetic utterance refers to what we are now considering tonight, Radium! Behold! the old boy beamed upon me!' Initially while Kelvin recognised that the radioactive material radium produced heat, he refused to believe that it produced the heat itself, and rather argued that the radium must have gained from an external source the heat that it subsequently emitted. However, soon afterwards he privately accepted that the discovery of radium had made some of his conclusions regarding secular cooling of the Earth difficult, but he never made this view publicly known. Perhaps it would have been best if he had slept through all of Rutherford's lecture, but there is little doubt that he saw the dawn of radioactivity herald the demise of his own geochronology.

12 Oceanic salination reconsidered

Sometime in July 1897 two men boarded the yacht *Marling-Spike* carrying with them a range of tubes, bottles and nets and other collecting paraphernalia. They set sail from the Bullock Harbour, close to the small picturesque village of Dalkey, ten miles south of Dublin, and steered the yacht southwards into Killiney Bay. They would have passed through Dalkey Sound with the low Dalkey Island on their left, on which was built a Martello tower. This robust, circular construction was just one of many around the Irish coastline built in the first decades of the nineteenth century as a first line of defence against a Napoleonic invasion. The advance never came and today the tower offers scant shelter against the frequent winter storms for the feral goats that eke out a precarious existence there.

On its landward side, Killiney Bay is dominated by a granite crag, topped with a monumental folly that overlooks the fine housing belonging to professional gentlemen and their families. Living on roads with names such as Sorrento Terrace and Vico Road, many are unaware of the similarity of their view to that at the Bay of Naples. The only geological feature missing is the dominant volcano; here the residents can reflect instead on the nature of the distant Great Sugar Loaf Mountain to the southwest. This is a mere pimple when compared with Vesuvius, and is not even volcanic in origin: composed of Cambrian quartzites, it has weathered to a fine conical shape, which has often been misinterpreted as a volcano. Back in 1897, a black trail of commuters would have walked daily down the hill to the railway, where they would have caught the northbound train to Dublin that began its journey from the holiday resort of Bray to the south. This railway line is perched somewhat precariously on boulder clays deposited on the retreat of various ice masses originating in the Irish hinterland, and the ice from Scotland that carried the celebrated microgranite from Ailsa Craig as small microerratics.

The yacht glided past the local landmark of Whiterock, so named because of the pale granite juxtaposed against the darker Ordovician sediments into which it was intruded some 450 million years ago during the closing of the Iapetus Ocean – a forerunner of the present Atlantic. Nearby, bathers emerged from large unwieldy bathing machines and ran like crabs over the coarse shingle to take the waters. Other, less hardy souls walked along the beach towards the assorted sheds, some of which served as makeshift cafes. Perched on some distant rocks a large dark shag attempted to eat rather too many fish.

Not concerned with the dalliances of relaxing city folk, the two men began their search for microorganisms called coccoliths. One would imagine them to be marine biologists; but one would be very far from the truth. While both were academics employed by Trinity College Dublin, one was the assistant to the Professor of Botany, while the other was assistant to the grandly titled Professor of Natural and Experimental Philosophy (who would now be called the rather blander Professor of Physics in younger universities). The former was Henry Horatio Dixon and the latter John Joly (Figure 12.1), who was the following year appointed Professor of Geology at Trinity College. Dixon had to wait eight further years before he succeeded as Professor of Botany at the same institution. They met at university and became life-long friends. In 1888 they travelled widely together as young men on the Continent, and in particular enjoyed climbing in the Alps. They both became proficient yachtsmen and sailed frequently in the west of Ireland, and off Scotland and Norway. Soon after their Killiney trips Joly decided that *Marling-Spike* was too small, and with a bank balance swollen with his new professorial salary of £500 per annum purchased the larger craft *Woodcock*. They both served as Commissioners of Irish Lights, which afforded them the chance to sail Ireland's coastal waters during the annual tour of inspection. Joly was fascinated by the sea, and later carried out some work for the

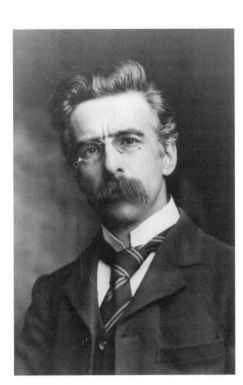

Figure 12.1   John Joly (1857–1933) in May 1901, shortly after the publication of his paper on sodium and the age of the Earth (Geological Museum, Trinity College Dublin).

Admiralty on signalling and safety at sea. During the First World War he bombarded the government with inventions that he felt would help the war effort.

On Joly's death Dixon inherited his house *Somerset*, which was set in a leafy mature southside suburb in Dublin, and moved his family in. They did not have to move far – the Dixons together with their three sons lived across the road. Although Dixon outlived his friend by 20 years, in death they were reunited, and now share a grave, together with Dixon's wife Dorothea, in Dublin's Mount Jerome Cemetery, a now rather decayed Victorian necrological park.

JOHN JOLY

Two questions need to be raised at this stage: what part do these yachtsmen play in the story of the age of the Earth, and what is the significance of their Killiney Bay trip to the present story? To answer the first part of the question first, Dixon contributed very little, certainly nothing in print, to the debate on the age of the Earth, but Joly contributed a great deal. He was to determine the age of the Earth by various methods, and his ideas on the subject were of considerable influence for nearly thirty years until his death in 1933.

Their expeditions to collect coccoliths in 1897 were instrumental in Joly's devising his earliest method of determining the age of the Earth, in which he examined the sodium content of the oceans. It is probable that as Dixon and Joly examined their catch, and then relaxed on deck, the conversation would have turned to other scientific problems. After all, it was these two men who cracked the sticky botanical question of how sap gets to the top of even the tallest trees. They showed in 1895 that the transpiration (or water loss) from leaves in plants sets up a pressure gradient in leaf and plant conductive cells that pulls the sap upwards. They were very close, shared holidays to the Alps, and certainly many of their ideas published separately would have been discussed and dissected.

Who was Joly? He was a man who plays an often central role in this story, and in many others, but whose contributions are now largely forgotten both in Ireland as well as in scientific circles, except to a few historians of science.

He was born on 1 November 1857 in Hollywood House (the Rectory), Bracknagh, County Offaly, the third and youngest son of the Reverend John Plunket Joly (1826–1858) and Julia Anna Maria Georgina née Comtesse de Lusi. The Joly family originated in France, but came to Ireland from Belgium in the 1760s. Joly's great-grandfather served as butler to the Duke of Leinster who gave the living of Clonsast parish to the family. After the untimely death of his father the Jolys moved to Dublin where John received his education at the celebrated Rathmines School run by the Reverend Charles William Benson. Benson, who was a keen ornithologist and author of a noteworthy slim volume *Our Irish Song-birds*, offered a liberal education, and encouraged individuality. Joly displayed a keen interest in and curiosity about science and continually tinkered with equipment, both laboratorial and that invented by himself. It was a trait, or perhaps self-training, that was to stand him in good stead for his later career path. At school he was popular, and earned the nickname 'The Professor' for his scientific tinkerings. In 1876 after a short period recovering in the south of France from a serious illness, he entered Trinity College Dublin, where he remained for the rest of his life. He followed courses in classics and modern literature but later concentrated on engineering, gaining the degree of Bachelor of Engineering in 1882.

John Joly spent all of his academic and professional life working in Trinity College, during which time he wrote 269 scientific papers and several books. Initially he was employed to teach and to assist the Professor of Civil Engineering, which he did from 1882 until 1891. As part of this research, apparatus had to be invented and built by Joly himself. Among his first pieces were a new photometer and a hydrostatic balance. He also developed an interest in mineralogy, and began to accumulate a large collection of some very fine Irish, Continental and American mineral specimens. He invented the steam calorimeter for measuring the specific heat of minerals, and this piece of equipment later played an important role in the kinetic theory of gases. In 1891, Joly was appointed assistant to George Francis Fitzgerald (Professor of Natural and Experimental Philosophy). During the following six years he invented and developed a mercury–glycerine barometer and an electrolyte ampere-meter, and published on the specific heat of gases. Joly's work was considered very important by his contemporaries and he was elected a Fellow of the Royal Society (of London) in 1892. This body was, and remains, the premier scientific organisation in the United Kingdom. If one examines his publications for 1895 and 1896 one can immediately appreciate the broad range of interests of the man: he wrote on the ascent of sap, on heat, and on gravitation, and he also published on his new method of colour photography. All in all, he had an incisive mind and was able to deal with complex scientific issues quickly, and, perhaps more importantly, was dexterous enough to devise and manufacture much of his own experimental apparatus. In 1897, he succeeded William Johnston Sollas in the Chair of Geology and Mineralogy, a position he retained until his death in 1933. Essentially Joly was a physicist and not a geologist and unlike his predecessors published little about the geology of Ireland. He carried out much research on minerals, but it was his work on radioactivity and radium that was his most important.

What was Joly like as a man? In adulthood Joly was a distinctive and unforgettable man; he was tall, with hair swept off his forehead, a bushy moustache, and pince-nez (early spectacles minus arms) perched on his nose. He spoke with what was considered to be a foreign accent, but in reality the rolled r's were simply employed to conceal a slight lisp that had afflicted him since boyhood. For recreation he travelled, read, collected works of art and maintained a good garden – although he succeeded in burning down his greenhouse, perhaps as a result of some failed experiment. Above all he enjoyed the company of his colleagues, and especially that of the Dixon family. One of the Dixons' sons bore his name and another was Joly's godson.

As an Irishman in the first two decades of the twentieth century Joly must have been affected by the local agitation that preceded Ireland's gaining independence from British rule in 1922. In common with many Irish Protestants at that time Joly was a Loyalist, which however did not stop him feeling proud to be Irish. It is a situation that today is unfortunately not understood by some elements of Irish society. During Easter week 1916 a group of men and women led by the poet Padraig Pearse occupied a number of landmark buildings in Dublin's centre, and declared an Irish Republic. Quickly the authorities in Trinity College closed up the university and, using the members of the College's Officers Training Corps and other academics, began to patrol the perimeter of the 40-acre city-centre campus. John Joly volunteered for duty, and spent four days in occupation. Although he did not fire a shot in anger, he did undertake dangerous forays from the campus on his bicycle to purchase cigarettes for the soldiers. After a week of fierce fighting the insurgents were forced to surrender, and were then rapidly executed. This turned public opinion, and they were seen as martyrs. Today most of the major railway stations in Ireland are named after one of the leaders of the 1916 rebellion. For their efforts in preserving the college from damage and potential destruction, a number of academics were awarded silver cups by the local business community. Joly's assistant, the shy palaeontologist Louis Bouvier Smyth (1883–1952), who succeeded him in the Chair of Geology in 1934, was given a ceremonial sword! He promptly lodged it in the Geological Museum but unfortunately, along with Samuel Haughton's slippers, it is now lost. It was to be another six years before Ireland gained independence, and Joly unlike many others chose to remain in Dublin and to throw his hat in with the new government. In fact Trinity College retained much of its ethos and continued at ceremonial dinners to toast the health of the King of England for at least a decade afterwards.

Joly had a high international reputation. He served as President of Section C (Geology) when the British Association for the Advancement of Science visited Dublin in 1908 – this body continues to meet annually to discuss topical scientific matters, and in 2005 returned to Dublin. Joly received the Boyle Medal of the Royal Dublin Society in 1911, the Murchison Medal (of the Geological Society of London) in 1923, and a Royal Medal from the Royal Society of London in 1910 – probably the most eminent scientific society in the world. He became a Fellow of Trinity College in 1919, and was President of the Royal Dublin Society between 1929 and 1932. Honorary degrees were conferred on him from the National University of Ireland, the University of Cambridge and the University of Michigan, which he visited as part of an British educational delegation sent to observe and report on the American higher educational system.

Joly was by all accounts a very popular man, loved and respected by many. On his death, friends contributed over £1,700, which was a considerable sum in 1934, to a memorial fund that is still used to promote an annual lecture in the University of Dublin. In addition his name and memory are perpetuated by the Joly Geological Society, the student geological association founded in 1960, in the same institution. Forty years after his death a crater on Mars was named after him, which is appropriate in view of his work published in 1897 on the nature and origin of Martian canals.

JOLY ON THE AGE OF THE EARTH

Joly's first documented thoughts on the antiquity of the Earth were in written verses penned on 28 August 1886 in the Wicklow Mountains, south of Dublin. These considered the age of the enigmatic trace fossil *Oldhamia antiqua* Forbes which is preserved in green and purple slates of Cambrian age. It occurs as faint fan-shaped markings, the origin of which still remains a mystery to palaeontologists. Joly suggested the fossil was a witness to the long, slow changes that had affected the Earth in a sonnet:

> *Is nothing left? Have all things passed thee by?*
> > *The stars are not thy stars! The aged hills*
> > *Are changed and bowed beneath repeated ills*
> *Of ice and snow, of river and of sky.*
> *The sea that raiseth now in agony*
> > *Is not thy sea. The stormy voice that fills*
> > *This gloom with man's remotest sorrow shrills*
> *The memory of the futurity!*
> > *We – promise of the ages! – Lift thine eyes,*
> > *And gazing on these tendrils intertwined*
> > *For Aeons in the shadows, recognize*
> > *In Hope and Joy, in heaven-seeking Mind,*
> > *In Faith, in Love, in Reason's potent spell*
> > *The visitants that bid a world farewell!*

Joly's first scientific foray into the subject of dating the Earth, or geo-chronology, came 13 years later in 1899 with the publication of his first, and probably most celebrated, if somewhat controversial paper, which expounded what became known as Joly's sodium method. Joly turned to the oceans for inspiration, which was not surprising. In the latter half of the nineteenth century a considerable volume of

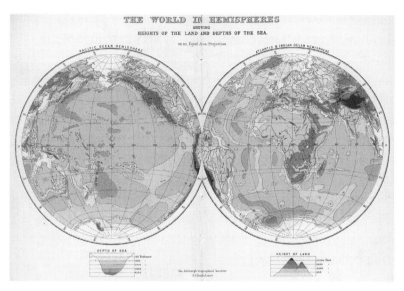

Figure 12.2   The Oceans (from A. Geikie, *Elementary Lessons in Physical Geography* (1907), Plate 1).

oceanographic research had been carried out, so that by the mid-1890s the physiographic features of many oceans were well known (Figure 12.2.) What Joly did was to examine the annual rate of sodium input into the oceans and by simple mathematics arrive at an estimate for the age of the Earth.

At this time Joly was unaware, as were all others enquiring into the age of the Earth, of the pioneering suggestions of the English astronomer Edmond Halley, whose work on water salinity we examined in Chapter 4. There are many examples where earlier theories and ideas have been forgotten, and remain in the pages of large tomes sitting on shelves high up in a musty library. Although published as a short note in the *Proceedings of the Royal Society*, which was widely taken by libraries, this did not stop Halley's ideas becoming lost in the ever-increasing volume of paper. Researchers working at the dawn of the twentieth century did not have the modern benefits of Georef and other on-line bibliographic databases which spew out numerous citations to scientific and geological papers depending on what key word one chooses to enter. Halley's note was only rediscovered in 1910 by George Ferdinand Becker (1847–1919), a geologist with the United States Geological Survey who had a background in physics and mathematics. Becker, who made numerous contributions to the geochronological debate, brought Halley's paper to the wider attention of the scientific community by penning a note in the premier American journal *Science*. Joly was aware of Mellard Reade's valuable paper published in the *Proceedings of the Liverpool Geological Society* in 1876 and the book that followed, which examined the volume of calcium sulphate and chloride in the oceans and derived dates of 25 million and 200 million years respectively, based upon their annual rate of accumulation. Later, Joly acknowledged these pioneering publications of Halley and Mellard Reade, but in 1915 noted without explanation that their schemes, unlike his, would not have produced reliable results.

On 17 May 1899 John Joly left his office in the Museum Building and walked through the College, and out on to the streets of Dublin. He turned up Kildare Street, passing the Kildare Street Club on his left, where some years before the cigar-smoking gentlemen had been startled by a cricket ball that smashed through a window, hit by W. G. Grace during a game against the Gentlemen of Ireland on the neighbouring College Park. On reaching his destination he turned left into a cobbled forecourt and walked towards the main entrance of Leinster House. Appropriately this formidable building had been built by the Duke of Leinster in 1743 in whose pay Joly's ancestor was. In 1899 it was the meeting house of the Royal Dublin Society, a body established in 1731 to promote science and agriculture: now it is the meeting place of the Irish Parliament or Dail. Joly entered the meeting hall, and, standing in front of the lectern, proceeded to read a paper entitled 'An estimate of the geological age of the Earth' to the assembled members.

The paper was rapidly published four months later in September in the Society's *Scientific Transactions*. Unusually, perhaps even uniquely for publications of this organisation, a second impression of the paper had to be produced in November 1899 as all the stocks of the original were distributed, and demand continued. This allowed Joly to correct a number of small errors that had appeared in the appendix of the first impression. The paper was also reprinted in North America in its entirety in the *Annual Report of the Smithsonian Institution* for 1899, all of which allowed for rapid and widespread dissemination of Joly's theory. This paper fired the imagination of both scientific and general audiences, and for perhaps a decade his 'sodium method' held sway amongst geochronologists. This work was partially influential in the discrediting of Lord Kelvin's chronology, which was beginning to fall from favour as the great man grew older. However, in due course even Joly's work at this time became discredited, and supplanted by the newer field of radiogeology.

Brilliant in its simplicity, Joly assumed that when formed, the oceans that bathed two-thirds of the Earth's surface were composed of fresh water. But now they are salty, with varing amounts of sodium, magnesium, and potassium salts and other materials such as calcium chloride. These materials must have been derived from minerals found in various rock types, which through aeons of erosion by rainwater and seawater became released and dissolved in these waters. Rivers, it was postulated, carried the bulk of the sodium into the oceans, and Joly's essential assumption was that the rates of denudation or erosion of the sodium-bearing rocks and the discharge of the rivers into the oceans had remained uniformly constant over geological time. So too the volume of sodium carried each year. This uniformitarian stance was one of the fundamental tenets of Joly's paper. The age of the Earth was derived by the simple formula:

$$\frac{\textit{Mass of sodium in the ocean}}{\textit{Rate of annual sodium input}} = \textit{The age of the Earth}$$

To derive figures for this equation Joly turned to the oceanographic and fluvatile findings published by Sir John Murray (1841–1914), a Scottish naturalist, in the 1880s in the *Scottish Geographical Magazine*. Murray focused on 19 rivers, including the Amazon, and calculated the annual volume of water and mass of various mineral salts and other materials carried by them into the oceans. Using results from the celebrated oceanographic expedition of HMS *Challenger*, Joly noted that the oceans contained 3.5% by mass of various salts of which sodium chloride [NaCl] ($35,990 \times 10^{12}$ tons), magnesium chloride [$MgCl_2$] ($5,034 \times 10^{12}$ tons) and magnesium sulphate [$MgSO_4$] ($2,192 \times 10^{12}$ tons) were the most abundant. Sodium constitutes just under 40% of sodium chloride, which meant that the mass of the element in the oceans was $14,151 \times 10^{12}$ tons. However, the rivers carry not only sodium chloride, but also lesser volumes of sodium sulphate [$NaSO_4$], sodium nitrate [$NaNO_3$] and sodium chloride [NaCl], which Joly showed, based on Murray's figures, contributed 157,267,544 tons of sodium annually into the oceans. When he put these figures into the equation above he got:

$$\frac{14,151,000,000,000,000}{157,267,544 \; tons \; per \; year} \; tons = 90,000,000 \; years$$

He concluded therefore that the age of the Earth was approximately 90 million years. With minor adjustments he widened this date to between 90 and 100 million years.

Reaction to Joly's paper began to appear in the scientific press within six months of its publication. Review articles were published in several journals, including the *American Journal of Science*, *Nature* and the *Geological Magazine*. Over the next decade a considerable number of papers discussed Joly's sodium method, and reaction was somewhat polarised: some authors were in broad agreement with his ideas, whereas others raised a number of objections. The Reverend Osmund Fisher penned the review that appeared in the *Geological Magazine*, and it was by far the most testing. Fisher was a combative character but well respected both as a geophysicist and as a cleric, in which role he ministered to the needs of his congregation at the

Church of the Assumption in the village of Harlton near Cambridge and served the university as a Chaplain of Jesus College. Like many a country cleric at that time, Fisher was able to indulge his passion for physics and geology, and was well versed in the scientific debates of his time. Regarding geochronology, he was familiar with the arguments proposed both in Europe and America – he had earlier, in 1893, reviewed Clarence King's work in the United States, which focused on cooling rates of igneous rocks to derive a figure for the age of the Earth. In his review of Joly's paper, Fisher argued that the processes invoked by Joly were not uniform throughout geological time. Additionally he suggested that Joly's figures for the volume of sodium delivered into the oceans by rivers might be at fault, and moreover was at pains to point out that Joly did not take into account the effect of 'fossil sea water' which Fisher noted was present trapped in sediments and elsewhere. One major area of contention that kept on recurring was that pertaining to the volume of recycled sodium. Joly estimated that 10% of the sodium chloride carried down in the rivers came from rainwater and thus was recycled. Fisher felt that this percentage was too high. William Ackroyd, the public analyst for Halifax, agreed, and so began a public debate played out in the pages of *Chemical News* and *Geological Magazine*, where the combatants became increasingly aggressive with each communication. Ackroyd even went as far as accusing Joly of avoiding and leaving unanswered his arguments. It must be suspected that Ackroyd did not fully understand the intricacies of Joly's lines of thought.

Not all reviews of the 1899 paper were negative, however, and Joly found some useful allies. One was Sollas, by then Professor of Geology at Oxford, and a major influence in British geological circles. Sollas, on pure scientific reasoning, and not on sentimentality or association, sided with Joly, stating in his 1899 presidential address to Section C of the British Association for the Advancement of Science meeting at Bradford that 'there is no serious flaw in the method, and Professor Joly's treatment of the subject is admirable in every way'. Sollas did, however, question the reliability of the data concerning the river discharge of sodium, and also queried the effects of geothermal cooling over time. Joly also delivered a paper to the same meeting, and his conclusions on the sodium method and its use in dating the Earth must have created quite a stir as his report was ordered by the General Committee to be published *in extenso*. Later papers by Sollas in 1909, by Becker in 1910 and Frank Wigglesworth Clarke (1883–1931), the chief chemist to the United States Geological Survey, in the same year laid minor criticisms at Joly's door. Sollas recalculated the annual discharge of the rivers from which he derived a date of 78 million years, but suggested the age of the Earth lay within the range 80–150 million years. Clarke, whose important work was in the compilation of extensive volumes of geochemical data, examined the rate of removal of sodium from the landmass and arrived at a figure of 80 million years, while Becker suggested that Joly's figure of 10% for the contribution of sodium recycled from the atmosphere was too high and that the volume was closer to 6%.

Joly's responses to the reviews of his work strongly reinforced his uniformitarian principle. However, he did accept that his estimate of the role of rainwater in providing sodium chloride might have been overestimated and might require further experimental work. With respect to fossil seawater, Joly stated that it could only have contributed 0.9% of oceanic sodium chloride and, as such, was negligible. His response to Sollas was that 'there is much reason to believe that the nineteen rivers ... afford an approximation as to what the world's rivers yield'. Indeed, he stated in 1911 that the findings of Sollas, Becker and Clarke, together with his own, gave concurrent results of *circa* 100 million years, and proudly anticipated that this determination would not be 'seriously challenged in the future'.

Joly went further in defence of his ideas in that he devised various experiments which he hoped would generate acceptance of some of the theoretical assumptions made in his 1899 paper. In one of these he devised a fractionating rain-gauge, which he hoped would allow him to collect rainwater over incremental time periods. Subsequent analysis of the amount of dissolved sodium chloride in this rainwater would allow him to quantify the volume of sodium in the oceans from this source. While demonstrating how his rain-gauge would operate in theory, there is no evidence to suggest that he actually built it, or, if so, that he put it to effective use. He also examined the rate of solution of various igneous materials in fresh and salt water and showed that of the four tested (basalt, orthoclase, obsidian and hornblende) the basalt from the Giant's Causeway in County Antrim dissolved more readily than the others, and that salt water was a more effective solvent than fresh water. Not surprisingly the obsidian, a volcanic glass, proved the most resistant to solution. Joly noted that his results for the rates of denudation were far lower than those demonstrated by field study and argued that additional factors such as organic acids, wetting and drying, and other erosive processes were more important than solution of rocks by water. Nevertheless he made an allowance for the solvent action of the ocean by reducing his age estimate by a few million years to 96 million.

After the initial peak of interest that followed closely from his 1899 paper, many of Joly's subsequent papers on the subject were simply reports of lectures or reiterations of the original theory. In 1915 with the publication of Joly's series of essays *Birth-time of the World*, which included readable and entertaining pieces on the colours of Alpine flowers and on skating, as well as a exposition on the antiquity of the Earth, there followed short-term, renewed interest in the sodium method – but this interest did not last. The theory was finally consigned to the scientific scrap-heap by several leading geologists on both sides of the Atlantic, such as the petrologist Alfred Harker (1859–1939) in 1914, John Walter Gregory (1864–1932) in 1921, Arthur Holmes (1890–1965) in 1926 in Britain, and Joseph Barrell in 1917 and Thomas Chrowder Chamberlin in 1922 in the United States. In damning words, Holmes rejected it as 'worthless'. Earlier in 1913 Holmes had cogently reasoned that the rocks undergoing erosion would have had to lose more sodium into the oceans than they ever contained for Joly's figures to add up. Holmes and Chamberlin, as will be seen later, laid the platform, built up and largely concluded the next method of absolute geochronological dating. The fickle scientific community had eyes only for the new theories based around radio-activity. Paradoxically, as we shall see, Joly also carried out much useful research in this developing area, but he himself could never consign his sodium method to the waste basket. Just three years before he died, Joly accepted some major modifications suggested by the American Alfred Church Lane (1863–1948) in 1929 which pointed at a figure of 300 million years for the method, but he was not prepared to concede to the longer timescales then in vogue. In the light of the well-advanced objections to his 'sodium method' and the findings from other methodologies, it is somewhat surprising that Joly did not accept that it yielded erroneously low age estimates. He obstinately held to his view, and did not acknowledge the more plausible and worthy conclusion of this work, that it probably measured the age of the oceans.

Joly's method assumed that the oceans formed at the same time as, or soon after, the formation of the Earth, so as to make little difference to his age determination. Given the nature of the primordial and early Earth this cannot have been true, and we now know that condensation of the oceanic waters from emitted gases took many millions of years, and that they are older than even Joly's method might suggest. What Joly's method actually equates closely to is the residence time of sodium in the oceans. It is now recognised that sodium spends anything between 70 and 100 million years swilling around in seawater before it becomes locked again in sediments which may in turn become lithified into rock. These may eventually be subjected to uplift and erosion, which would release the sodium once again into the oceans.

It is interesting to note that, on the basis of research published in 1998 by Paul Knauth in *Nature* on the volume of chloride contained in brines found in deep-seated groundwater, it is now thought that the earliest ocean was 1.5 to 2 times saltier than that of today. The sodium chloride was formed from the combination of chloride ions derived from outgassed volatiles and sodium ions leached from rocks. Through geological time the oceans have not become progressively saltier as was Joly's contention, but attained their high levels of salinity early in their life. This finding alone invalidates the basis of Joly's 1899 paper.

13 Radioactivity: invisible geochronometers

While most people can name the inventor of the television – the Scottish scientist John Logie Baird (1888–1946) – few of us can remember who invented the cathode ray tube, a fundamental part of the television without which it just would not work. For the first cathode ray tube, we can thank Sir William Crookes (1832–1919), an affluent chemist who had established his own private laboratory in London. The kind now used in televisions is a variant designed by Karl Ferdinand Braun (1850–1918), the German physicist whose name is now on many domestic electrical appliances. While working with a Crookes Tube in his laboratory at the University of Würzburg in Germany in 1895, the physicist Wilhelm Röntgen (1845–1923) made a remarkable discovery. When he turned the cathode tube on he discovered that barium platinocyanide sitting on a shelf on the other side of the room started to glow slightly. He then removed the chemical to the next room, and it still glowed. The explanation suddenly occurred to him: while the tube was producing cathode rays it was also producing another invisible ray that had the ability to pass through thin sheets of metal, which the cathode rays could not penetrate. As he had no clue what the rays were or what formed them he simply called them 'X-rays'. He then experimented with their properties and produced an image of his wife's hand by placing it on a photographic plate and exposing it to the X-rays. While the traces of her flesh could be detected, her bones as well as some rings were easily seen. Immediately a medical use was found for these new invisible rays. When others in the early twentieth century referred to them as Röntgen Rays his name became universally adopted.

Shortly after Röntgen's ground-breaking discovery a different source of invisible radiation was discovered. These rays were different from X-rays, but nevertheless were later shown to be useful in medicine. Their discovery changed the course of geological and geo-chronological study. This was radioactivity, a phenomenon discov-ered by Antoine Henri Becquerel (1852–1908) in February 1896 while he was experimenting with compounds of uranium. In a darkened room he had placed some potassium uranyl sulfate into a sealed box and by chance had rested an unexposed photographic plate beside it. When the plate was developed it was foggy and demonstrated that some radiation had passed through the box to affect the plate. He then discovered that the path of the emissions could be deflected by a magnetic field, thus showing that they were composed of tiny parti-cles. Within several years three types of rays had been described: alpha rays (α), which were weak and were absorbed by a thin metal foil; beta rays (β), which were more penetrative and easily deflected in a magnetic field; and gamma rays (γ), which were highly pene-trating and were not deflectable. Today these rays are measured by the eponymous detector invented in 1913 by Hans Geiger (1882–1945).

The search was now on for elements that produced such emis-sions, which Marie Curie (1867–1934) called 'radioactivity'. Obviously uranium was one such element, an element that had been known since its isolation from pitchblende by the German chemist Martin Klaproth (1743–1817) in 1789. In 1898 Marie Curie and her husband Pierre (1859–1906) isolated a new radioactive element from pitch-blende, which they called polonium, after the country of her birth, but they realised that another as yet unisolated element must be present and they called this radium. Eventually in 1902 after years of refining tonnes of pitchblende purchased from Czech mines they managed to isolate one-tenth of a gram of radium. Marie also demonstrated that thorium, which had been discovered by the English chemist Smithson Tennant (1761–1815), was radioactive. In 1899 actinium was discov-ered and named by André Louis Debierne (1874–1949), an assistant of the Curies, and researchers realised that some forms of lead showed radioactive properties. Some discoveries proved to be episodes of false hope: 'Hibernium' was named by John Joly in 1922 but was later discovered to be in fact samarium.

A SELF-GENERATING HEAT SOURCE: HEATING UP THE COOLING EARTH

So how did geological research benefit from the discoveries of radioactive elements? In several ways. By the turn of the twentieth century the geological community was in turmoil. The evolutionary ideas and the chronology needed for them were at variance with the chronology based on Thomson's cooling Earth model, and the sedimentation and oceanic salination models produced figures of approximately 100 million years that appeared to be too low for the palaeontologists too. There was no consensus and geologists were wading through arguments over theories that appeared to have little foundation. Suddenly radioactivity provided one solution to this problem: heat.

One day early in 1903 Pierre Curie, and his assistant Albert Laborde, noticed that the radium that he and his wife had so lovingly isolated appeared to be generating heat. To test this they placed in a sealed tube, fitted with a thermocouple to record any temperature rise, one gram of barium chloride contaminated with a small amount of radium, and into another similar tube placed a gram of pure barium chloride. Some time later they found that the radium-bearing tube was 1.5 degrees Centigrade hotter than the comparative tube, and subsequently they worked out that the heat produced daily by one gram of radium could melt one gram of ice. Radium was a source of heat, and they announced this irrefutable fact to the scientific world on 16 March 1903. In trying to understand the source of the heat Curie and Laborde suggested that it was either generated during the breakdown of the radium atom, or that it might have been absorbed by the radium from elsewhere. Joseph John Thomson (1856–1940) (no relation to Lord Kelvin) in 1903 suggested that the heat might have been produced as the atom contracted. At much the same time Ernest Rutherford (Figure 13.1) and his assistant Frederick Soddy (1877–1956) discovered that a great deal of energy was produced as

Figure 13.1   Ernest Rutherford, Lord Rutherford of Nelson (1871–1937), a pioneer of the study of radioactivity, recipient of the Nobel Prize for Chemistry in 1908. Photograph 1932 on the occasion of the announcement of the splitting of the atom by Ernest Walton and John Cockcroft. Courtesy of the School of Physics, Trinity College Dublin.

alpha particles were released, and shortly afterwards Rutherford and another of his assistants Howard T. Barnes (1873–1950) realised that the amount of heat generated was proportional to the number of alpha-ray emissions.

Radium was producing heat and energy in the laboratory, and it was not a quantum leap to realise that radium would also be generating heat in its natural environment within the Earth. Here was a source of some, if not all of the Earth's internal heat, and in 1903 John Joly audaciously suggested that this would render William Thomson's cooling Earth chronology invalid. Gone was the chronological impediment to Darwin! Immediately geologists realised the implications for their science, and Joly was one of the first to carry out wide-ranging measurements of the levels of radioactivity in terrestrial materials. He launched into this research with gusto, and produced a pioneering book *Radioactivity and Geology* on the subject in 1909 in which he documented the quantities of radium and thorium in a wide range of terrestrial materials, rocks and minerals as well as seawater. Most materials contained some traces of radioactive elements. Paradoxically he did not seem to realise the indirect effect that the new studies would have on his oceanic sodium chronology which too would soon be dead in the water. Nevertheless Joly's work on radium did have a beneficial side effect. With a Dublin doctor, Walter Clegg Stevenson (1876–1931), he established the Radium Institute at the Royal Dublin Society in 1914, which specialised in treating cancer patients. Their 'Dublin method' was the first to utilise radium emanation (radon) enclosed in hollow needles in the treatment of tumours, and this method is still used today in some procedures. Joly was very proud of this work. In the context of geochronology, however, his work with the Radium Institute is a minor, albeit interesting digression.

THE FARMER'S SON FROM NEW ZEALAND

Ernest Rutherford was probably the prime mover in the study of radioactivity. Second of twelve children, he was born on 30 August 1871 in the small town of Bridgewater, near Nelson in New Zealand, to recently arrived emigrants from England. For a time it seemed likely that he would follow his father into his trade of wheelwright, or possibly turn his hand to farming. However, the course of his life changed when the bright boy won a scholarship to Canterbury College, Christchurch, and from there another to Cambridge. He very nearly did not get the chance, but the student who was ranked first in the examination decided to remain at home and get married. On being told of his success Rutherford who was lifting potatoes at home is said to have exclaimed 'That's the last potato I'll dig!' In Cambridge he came under the influence of J. J. Thomson and there he began research on radioactivity, and the understanding of geology and the Earth's internal heat as it was known at the time was soon changed. He spent a brief interlude away from England when he took up a professorship at McGill University in Montreal, but by 1908 he was back, first at Manchester and then from 1919 in Cambridge again where he accepted the Chair of Physics in 1919. In the following year he took over as Director of the Cavendish Laboratory, succeeding his mentor J. J. Thomson. Rutherford was responsible for some major scientific discoveries but it was his work on radioactivity, and his investigations from 1909 into the internal structure of the atom, that were his most important. He visualised the atom as having a nucleus around which whizzed electrons: a structure similar, on a much smaller scale, to that of the Sun and its revolving attendant planets.

Rutherford published a landmark book in 1904 that was simply entitled *Radio-activity*, and this was instrumental in bringing the infant subject to a whole new audience of academics and students. The subject was progressing so fast that he had to produce a second edition of the book the following year, and it was 181 pages longer than the first. In 1908 he received the Nobel Prize for Chemistry; twenty-three years later he was raised to the peerage and thereafter was styled 'Baron Rutherford of Nelson'.

RADIOACTIVITY AND THE DISINTEGRATION SEQUENCES

In 1902 Rutherford and Soddy had a brilliant insight into the nature of radioactivity: they suggested that during radioactive decay of an element it becomes transformed into another, and these transformations became known as 'radioactive decay series'. Within a year the decay series of uranium, thorium, radium and actinium were known. The sequence for thorium was:

Thorium (the initial starting product) → thorium X → thorium-emanation (a gas now called thoron) → thorium-excited I → thorium-excited II → unknown (the resultant material)

and each step was accompanied by emissions of different rays at varying intensities. Rutherford also was the first to arrive at the concept of the 'half-life' or 'half-transformation' as he put it. This is the time period in which the activity of a radioactive element drops to half its original value. For example, in the first step of the uranium decay sequence between uranium and uranium X 1, the next element in the line of descent, the half-life is 4.51×10^9 years (a very long time) and it is *always* this long. In some other radioactive atoms the half-life between two adjacent steps of the series may be as low as 0.019 seconds.

By 1908 some additional steps had been inserted into the decay sequences (Figure 13.2), and later still more were added as further research was carried out. In 1910 Soddy wondered if there existed varieties of the radioactive elements, varieties that in 1913 he called 'isotopes'. An isotope is a species of an atom that has the same atomic number but different number of neutrons in its nucleus. All nucleii contain protons, which have a positive charge, and neutrons, which have no charge. The atomic number of an atom is the number of positive units in its nucleus. Soon after Soddy's speculations it was discovered that there were a whole host of radioactive isotopes. Uranium alone has several isotopes: ^{234}uranium (^{234}U), ^{235}U and ^{238}U while lead (Pb) has ^{204}Pb, ^{206}Pb, ^{207}Pb and ^{208}Pb.

What is remarkable is that within a few years of the discovery of radioactivity so much was known about it. In 1903 a journal *Le Radium* was established in Paris and published by Buffon's old publisher Masson of Boulevard Saint-Germain. It was supported by a Comité Scientifique which numbered Becquerel, Madame Curie, Debierne and Rutherford among its members. It is therefore hardly surprising that it became the premier journal for students of radioactivity. The journal also carried advertisements, one of which offered one milligram of pure radium bromide for 400 francs and one gram of uranium salt for 1 franc (Figure 13.3). When we compare what was known of the decay sequences by the beginning of the First World War to what was known sixty years later, there is remarkable similarity. Today the decay series are slightly longer thanks to the insertion of some additional steps into the pioneers' schemes.

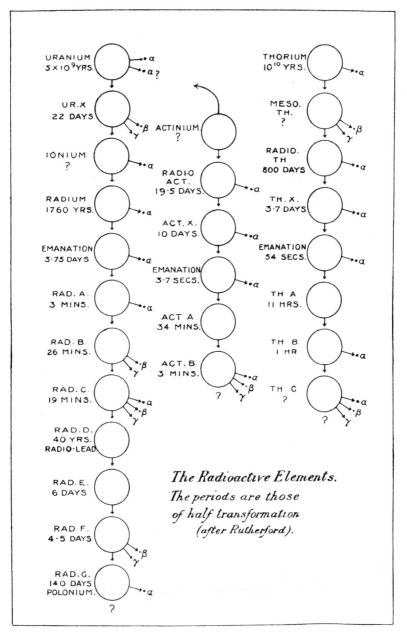

Figure 13.2   Radioactive decay series for uranium, thorium and actinium
(from J. Joly, *Radioactivity and Geology* (1909), Plate 2, facing p. 3).

Figure 13.3 Advertisement from the back cover of the journal *Le Radium* **5**, part 6 (July 1908) offering radium, polonium, actinium and uranium salts for sale (Geological Museum, Trinity College Dublin).

DECAY SERIES AND GEOCHRONOLOGY

One might wonder, then, what good these radioactive decay series could be in unravelling Earth history and chronology. This challenge was taken up by two young men working either side of the Atlantic, by a researcher at Yale University with the wonderful name Bertram Borden Boltwood (1870–1927) (one can imagine him gracing the croquet lawn at Blandings Castle), and by the Englishman, the Honourable Robert John Strutt (1875–1947), better known later as the fourth Baron Rayleigh. (Unusually and quite incidentally the Barony was first conferred on a woman, Robert's great-grandmother the daughter of the Duke of Leinster, and the title of Baron was then passed down the male line.)

In 1905 Strutt, who was working in Cambridge at the important Cavendish Laboratory, took up Rutherford's speculation that the gas helium was an end product of a decay series, and went further in suggesting that if the quantity of trapped gas in rocks could be determined then it might be possible to discover the age of the rock. Over the next five years he worked on this problem, analysing many different rocks and minerals. He said that the deposits of iron ore haematite at Frizington in Cumbria, which lay just above the Carboniferous limestone, were a minimum of 141 million years old, while fossilised sharks' teeth from Florida were 77 million years old. These determinations were not bad, when compared with present estimates. The iron ore is likely to be no more than 300 million years old while the Miocene teeth are probably between 5 million and 23 million years old. Strutt realised that he would have problems dating these materials accurately as it was highly probable that some of the helium as it formed was lost through seepage and leakage. Consequently by 1901 he considered that this line of research led up a cul-de-sac and so abandoned it.

At much the same time that Strutt was labouring in Cambridge, Boltwood thought that he had discovered a new element, ionium, but unfortunately for him it was later discovered to be a variety of thorium. However, his other findings received great attention. Having examined the uranium decay series in detail he worked out that the end product of the series was the stable element lead. Once lead was formed the series stopped. He then produced a momentous argument which went as follows: if one determined the ratio of the radioactive uranium to the stable lead in the ore, and knew the decay rate (half-life) of uranium to lead, then it would be possible to work out how old that ore was. This method could then be used to establish the age of any uranium/lead-bearing rocks in the Earth's crust and could give an age for the Earth itself.

So for clarification let us take a simple hypothetical example with invented figures. A mineral grain is analysed and found to contain 100 grams of radioactive element A and 300 grams of stable element B. The half-life of the A→B decay is 10 years. Originally when formed the grain must have contained 400 grams of element A, but after 10 years' transformation left 200 grams of element A and 200 grams of element B. In the next ten years half of the remaining element A is transformed into element B giving the ratio of 1:3 seen in the mineral, which therefore must be 20 years old.

On the basis of his argument Boltwood examined a series of mineral specimens and for these he obtained a suite of age determinations that ranged from 400 to 2,200 million years, results he announced in a paper in the *American Journal of Science* in 1907. These figures ten years earlier would have been considered fantastic, in the true meaning of the word, and Boltwood would have been considered unhinged. But in 1905, while the results were startling, the geological community sat up, took notice and began to agree.

Agreement was not universal, however, and some such as Joly argued otherwise. He questioned whether the decay rate of uranium has been constant throughout geological history as had been suggested. He said that this assumption was without strong basis and that the published results were based on derived radioactive products rather than radioactive parent products. Joly was not alone in his concerns. Becker also voiced unease with radiometric dates, as did the Canadian Committee on the Measurement of Geological Time

Figure 13.4 Arthur Holmes c. 1911. Reproduced by kind permission of the Geological Society of London.

by Atomic Disintegration, which reported that uranium as a whole decayed more rapidly in the past, and therefore the dates derived from it could be overestimated by 25%. The English geologist Arthur Holmes (Figure 13.4) argued in 1924 that at worst, errors of a few per cent might be determinable for what became known as the radiometric method. Holmes was right to be optimistic: the objections were soon negated by further research.

Strutt's and Boltwood's work combined to be among the most innovative and brilliant advances in the annals of geochronology. Strutt went on to a career in the Royal College of Science in London, while poor Boltwood ended up having a nervous breakdown brought on by the strain of overwork, and while recuperating on the Maine coast in August 1927, he became depressed and committed suicide. He was only 57 years old.

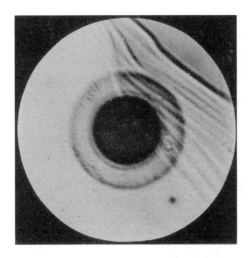

Figure 13.5 A single pleochroic halo developed in biotite in the Leinster granite from Garryellen, County Carlow, Ireland. The inner dark disc is due to radon (a gas derivative of radium); the succeeding inner ring is due to radium A alpha decay while the outer darker ring was produced by radium C alpha decay (from John Joly, 'Radiant Matter', *Scientific Proceedings of the Royal Dublin Society*, New Series **13**, part 6 (1911), Plate 3, Figure 4).

PLEOCHROIC HALOS: THE CHRONOLOGICAL RINGS

In 1907 Joly realised that small dark rings, or pleochroic halos as they came to be known, which he had observed in biotite in some granites were the products of radioactive decay in zircons enclosed by the biotite crystals (Figure 13.5). Previously it had been suggested that these were due to the presence of organic pigments in the minerals, but Strutt had earlier demonstrated the radioactive properties of zircon, to which Joly attributed the halos. The size of the halo was related to the type of radioactive decay product and the range of the rays produced and the intensity, he argued, was due to the duration of radioactive decay. He observed complex halos with distinctive inner and outer rings, or corona, in a greisen from Saxony in Germany, and in 1910 he and his research assistant Arnold Lockhart Fletcher (d. 1917), who was killed shortly afterwards near Rouen in France during the First World War, attributed the development of the outer rings to alpha rays of radium C and the inner rings to radium A. Subsequently Joly was able to distinguish halos produced by various radioactive sources including thorium, radium and uranium.

In 1913 Joly and Rutherford developed a unique methodology to date a rock on the basis of its pleochroic halos, which required knowledge of the mass of the nucleus of the halo and the number of alpha rays required to produce a certain intensity of halo. Using specimens from the Leinster Granite of Carlow in Ireland, Rutherford, working in Manchester, produced artificial halos in mica and measured the number of alpha rays required to produce the halos. Meanwhile in Dublin, Joly measured the mass of the nucleii and between them tabulated the age of thirty halos. These age estimates ranged from 20 to 470 million years. They concluded that the age of the Devonian was not less than 400 million years. However, a proviso read: 'that if the higher values of geological time are so found to be reliable, the discrepancy with estimates of the age of the ocean, based on the now well-ascertained facts of solvent denudation, raises difficulties which at present seem inexplicable'.

Three years later Joly measured halos in younger rocks from the Vosges, and later still he measured the radii of rare halos in the Tertiary granites of the Mourne Mountains. The radius of the latter were 7% smaller than those of the Leinster Granite, which themselves were 10% smaller than those he had recognised in Archean rocks, and in 1917 he concluded in triumph: 'It would seem as if we might determine a geological chronology on the dimensions of these halo-rings!' No sooner had he come to this exciting conclusion than he discovered small halos in Archean rocks from Norway which seemed to explode their chronological promise. Subsequently Joly's halo data were re-examined in 1927 by D. E. Kerr-Lawson of the University of Toronto who was unable to detect the differences in size Joly had noted in the different rock types. Shortly afterwards another student of pleichroic halos, Franz Lotze (1903–1971) of the University of Göttingen, noted anomalies in the rings caused by uranium and said that this invalidated Joly's ages derived from halos.

In 1922, Joly had reiterated his contention that radioactive decay rates were not constant throughout geological history. This was partially based on his observations of halos the characters of which were not consistent. In particular, the innermost rings produced by uranium in some halos were not consistent with the known ionisation curves of uranium alpha-particles. He suggested that the rings were caused by the faster decay of uranium in the past, or by the decay of a uranium isotope that was no longer present. Similarly, other isotopes might have been present which have now disappeared, and so the decay rates of these would be unknown. The fact that thorium halos of all ages were constant in size was difficult to explain. Arthur Holmes argued that the inconsistencies of the uranium halos were due not to time but to other factors including the presence of the recently discovered rare uranium daughter isotope actinium. Nevertheless he accepted that Joly's scheme of correlating halo radii with time would give a scale against which the ages of other halos could be determined.

PUTTING AGES ON THE GEOLOGICAL COLUMN
Boltwood's method is, in essence, the fundamental basis of radiometric dating, which ultimately allowed geologists to fix absolute dates to the geological column. Today geologists use one or a combination of various decay sequences (which begin with a radioactive 'parent' isotope and end with a stable 'daughter' isotope) to arrive at an absolute date for a particular rock (Table 13.1). The last three decay series have been developed in the past 25 years following improved analytical instrumentation.

In 1911 the only methods available to Arthur Holmes, who at the time was a young research student working in Strutt's laboratory in London, were the uranium → lead and the helium accumulation methods. Holmes recognised the value of the new methodologies as tools which he could use to provide absolute dates to rocks from known portions of the geological column. He wanted to provide an absolute timescale for geological history, and throughout his life he remained at the cutting edge of this research and was its prime mover and shaker.

Holmes' peregrinations
Holmes was born in 1890 in the village of Hebburn, near Newcastle in northeast England, but spent his childhood at the nearby town of

Table 13.1 *Radioactive decay series most usually used for geological dating.*

Uranium → lead	[^{238}U decays to ^{206}Pb with a half-life of 4,470 million years]
	[^{235}U decays to ^{207}Pb with a half-life of 704 million years]
Thorium → lead	[^{232}Th decays to ^{208}Pb with a half-life of 14,000 million years]
Potassium → argon	[^{40}K decays to ^{40}A with a half-life of 1,250 million years]
Rubidium → strontium	[^{87}Rb decays to ^{87}Sr with a half-life of 48,800 million years]
Samarium → neodymium	[^{147}Sm decays to ^{143}Nd with a half-life of 106,000 million years]
Rhenium → osmium	[^{187}Re decays to ^{187}Os with a half-life of 43,000 million years]
Lutetium → hafnium	[^{176}Lu decays to ^{176}Hf with a half-life of 35,900 million years]

Source: half-lives after G. Brent Dalrymple, *The Age of the Earth* (1991), p. 80.

Gateshead where his father ran an ironmongery and later was engaged in the insurance business. The young Arthur shone at school, and in his final year first encountered the works of Lord Kelvin and also read parts of *The Face of the Earth* by the German geologist Eduard Suess (1831–1914). The year 1907 saw him take up a scholarship at the Royal College of Science in South Kensington in London where he studied physics, but also took some courses in geology – a training that moulded his future work. He applied for a position as a petrologist, someone who specialises in the study of rocks, in the British Museum (Natural History) but was unsuccessful, losing out to William Campbell Smith (1887–1988) who proved to be a very able mineralogist and petrologist. Sensibly Holmes had more than one iron in the fire, and he was offered another scholarship worth £60 per annum to remain at the Royal College. Sixty pounds was not a great deal of money, but to an impoverished student it gave him a lifeline to extend his studies and commence research in earnest, and he was able to supplement this sum by writing book reviews for *The Times*. Entering Strutt's laboratory he was given the task of finding a better dating method than the helium accumulation method, which produced leaky results. It says something about Holmes' character that he had the ability to set aside his own research and look for better alternatives: pride stops many scientists taking similar brave steps. In 1911 he was offered a job in Mozambique with Memba Minerals Limited which hoped to exploit the copper and tin deposits in the country. The job was worth seven times his scholarship, so he went to Africa, but things did not progress smoothly. The sixty donkeys that had been purchased to work as carriers of equipment and personnel were riddled with disease and all but two had to be killed. Later one of the two remaining animals did not endear itself to one of the geologists: it bit his arm and he nearly lost it to an infection that set in. The expedition trekked inland approximately 250 miles and slowly passed over predominantly Precambrian rocks. This proved useful in a lateral way, as Holmes was able to collect samples of the mineral zircon, which he later dated, and the group also found thorium, but precious little else.

He returned to England in late 1911, and soon afterwards married a Gateshead girl. They were to have two children, but the elder died at the tragically young age of 5 in Burma, where his father had gone to work in 1920 for the inefficient and corrupt Yomah Oil Company. Prior to his Burmese sojourn Holmes spent time writing a short book, *The Age of the Earth*, which was published to critical acclaim in 1913, and working as a demonstrator at Imperial College (a rebranded Royal College for Science). Between 1922 and 1924 after returning from Burma he eked out a living as the proprietor of a shop in Newcastle selling exotic goods from India and the Far East. It must have been a trying and frustrating time for him as he yearned to return to academia and the age of the Earth problem.

Salvation came when he was offered the Chair of Geology at Durham in a new Geology Department. He threw himself into teaching and research and rapidly built a department that still retains a fine reputation. Not only did he return to geochronological research but he began work that culminated in his proposing that continental drift was driven by convection currents in the mantle of the Earth, which were themselves driven by the Earth's internal heat. He also turned his attention to the study of igneous rocks in Ardnamurchan in Scotland, and it was here in 1930 that he met Doris Livesey Reynolds (1899–1985), a lecturer in geology at University College London. Arthur was bowled over twice, firstly by the geology of the island, and secondly by Doris. Three years later she was appointed to the staff of the Geology Department at Durham and was installed in Arthur's office from where they then both worked. Their developing closeness raised eyebrows in the college, but following a lingering illness, Arthur's wife Maggie died. After a respectable period of nine months Doris and Arthur married, but out of the blue came a surprising and momentous decision by the university authorities: they would not renew Doris's contract. Following representations (by her husband) she was offered a year's extension, but Holmes resigned and the pair decamped to Edinburgh when in 1943 he was appointed Regius Professor of Geology. He remained in Scotland for the remainder of his university career, but moved to London on retirement in 1956 where he died in 1965.

The development of Holmes' geological timescale

From his very early research in 1911 he challenged the geochronological views of his older peers, and in his very first paper published in the *Proceedings of the Royal Society* that year (when he was only twenty-one) he gave an age of 1,640 million years for the Archean rocks of Ceylon based on the decay of uranium to lead in radioactive minerals that were frequently found in the mineral zircon. Up to that time this was the oldest date recorded for a rock. In the paper he reported the results of eight analyses; he had examined seventeen minerals but the findings from just over half were problematic. He arrived at a date for the Silurian/Ordovician of 430 million years and for the Devonian of 370 million years (Table 13.2).

Table 13.2 *Changing views of Arthur Holmes' Geological Timescale 1911 to 1960, compared with the 2004 timescale published by the International Commission on Stratigraphy. The dates refer to the date of the beginning of the geological units in millions of years, except where indicated by* [a] *where the date of the end of the unit is tabulated instead. The dates derived from helium are marked* [b]. *The 1947 figures are those of his 'B' listing. Where Holmes gave a range of dates the greatest figure has been listed.*

	1911	1915	1935	1947	1960	2004
Pleistocene	–	1.0	–	1	1	1.8
Pliocene	–	2.5	–	12	11	5.3
Miocene	–	6.3	38^b	26	25	23.0
Oligocene	–	8.4^b	–	38	40	33.9
Eocene	–	30.8^{ab}	37^b	58	60	55.8
Paleocene	–	–	–	–	70	65.5
Cretaceous	–	–	60^a	127	135	145.5
Jurassic	–	–	128^a	152	180	199.6
Triassic	–	–	170^b	182	225	251.0
Permian	–	42.7^b	205	203	270	299.0
Carboniferous	340	320^a	196^a	255	350	359.2
Devonian	370	380	278^a	313	400	416.0
Silurian	430	–	–	350	440	443.7
Ordovician	–	–	380^a	430	500	488.3
Cambrian	–	–	455^a	510	600	542.0
Precambrian	1,640	–	–	–	–	–
Upper Precambrian	–	307^b	600	–	–	–
Middle Precambrian	–	1,200	–	–	–	–
Lower Precambrian	–	1,500	–	–	–	–

Holmes' 1911 paper can be regarded as one of the greatest pieces of geological literature ever published. The paper had appeared only four years after the death of Kelvin, and while the enormity of the date and his work did not please everyone in the geological community, it was not long before the general consensus was that Holmes was *right*. The paper was the first in a line that eventually produced an absolute timescale for the geological column. Throughout his life Holmes continued to refine his geological timescale as his own research or that of others produced more accurate dates (Table 13.2).

In 1915 he calculated the ages of more minerals taken from a wider stratigraphical range than he had attempted in 1911, but there were still gaps in the dating scheme: he had no Mesozoic data and a date for the Silurian was absent. By 1935 the coverage was better still and he noted that the Earth was no less than 1,900 million years old, and that it seemed likely that the Earth and the Solar System could be the same age, which was approximately 2,000 to 3,000 million years. By his 1947 paper he had produced a timescale that for all intents and purposes is very similar to the 2004 version. In 1947 he produced two scales: the A scale defined the maximum age limit of the geological unit and the B scale the lower age limit of the unit (Figure 13.6). The B scale became more widely quoted subsequently. The precision of dates in the 1947 scheme had been improved since the discovery that uranium formed isotopes one of which decayed to ^{207}Pb. Prior to this, ^{207}Pb had been thought not to have been formed by radioactive decay, but if this was so, was there any lead isotope not formed by radioactive decay? If there was then all the previous results using the uranium–lead method would be inaccurate, and one could only determine an accurate age if one knew the original proportions of the uranium isotopes. In stepped Alfred Otto Carl Nier (1911–1994), known to all as Al, who was working at Harvard at the time. He modified a mass spectrometer, a machine used to measure the mass of atoms, improving its precision, and this allowed him in 1939 to determine the ratio of the two uranium isotopes. He said in 1939 that if one examined a lead ore, one found that a proportion of it was lead

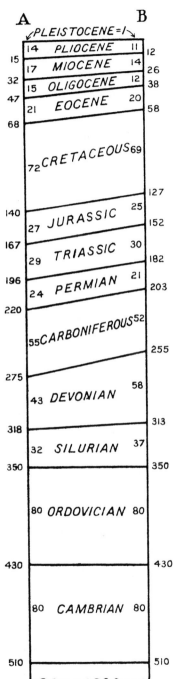

Figure 13.6   Holmes' 1947 geological timescale. The figures in the boxes refer to the duration of the geological periods in millions of years; those outside refer to cumulative ages before the present. Holmes preferred Scale B (from Arthur Holmes, *Transactions of the Geological Society of Glasgow* **31**, part 1 (1947), p. 145). (Geological Museum, Trinity College Dublin.)

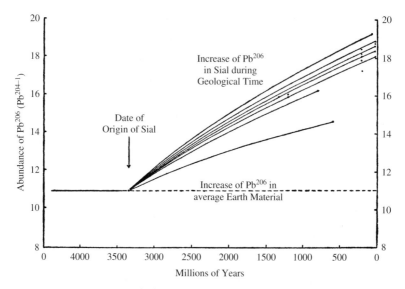

Figure 13.7   Graph showing seven isochrons converging on a point 'E'
indicating an age of 3,400 million years (from Arthur Holmes, *Geological
Magazine* **84** (1947), p. 124). (Geological Museum, Trinity College
Dublin.)

formed over time by the radioactive decay of uranium and thorium
and that the rest was what he called 'Primeval Lead': lead that was
present at the origin of the Earth. This was a major breakthrough
that should not be understated, as it allowed others to make serious
refinements to our understanding of our planet's age. Arthur Holmes
took Nier's results on board and in 1946 established a dating procedure
based on this new information, which took into account the different
decay times of the two uranium isotopes. If the proportions of the two
uranium-to-lead systems are plotted on to a graph, the points form a
straight line, called an isochron, and the slope of this line represents
the age of the Earth (Figure 13.7). The following year Holmes, using
the method, calculated that the Earth was 3,350 million years old.
Unknown to him some of his ideas had been considered earlier by the
Russian Erik Karlovich Gerling (1904–1985) in 1942, and the method
was developed almost simultaneously in Germany by Friedrich (Fritz)
Georg Houtermans (1903–1966). Houtermans was imprisoned in

Moscow just before the Second World War and under some pressure confessed to being a spy. He was repatriated to Germany and eventually became a professor in Bern. The dating methodology established at this time is now known as the Holmes–Houtermans Model. In 1956 Holmes reported a date of 3,450 million years for a rock collected at Monarch Reef, East Champ d'Or in the famous Witwatsrand district of South Africa. In his 1960 timescale Holmes corrected a number of errors that he had discovered in his 1947 scale. A number of rocks collected from the Appalachian Mountains in the eastern United States were found to be Devonian and not Ordovician as previously thought. He also pushed the base of the Cambrian back by 90 million years to 600 million years: today this has been reduced to 542.5 million years. However, setting aside this minor adjustment of 57.5 million years it is a remarkable testament to the labours, accuracy and tenacity of Arthur Holmes that his timescale of 1960 is so similar to that published in 2004.

Holmes was a giant in the world of radiometric dating and its application to the geological time and the geological column. He served as a member of two committees of the National Research Council in Washington: the 'Subsidiary Committee on the Age of the Earth', which reported in 1931 (Holmes contributed the lion's share of the report); and the 'Committee on the Determination of Geological Time' which reported in 1934. In 1956 he received both the Wollaston Medal from the Geological Society of London and the Penrose Medal from the Geological Society of America, the highest awards both bodies could make, and he shared the important Vetlesen Prize in 1964. However, as Cherry Lewis has documented, Alfred Nier bestowed a far greater compliment on Holmes when he described his as the 'Father' of geological timescales. A simple but fitting tribute to a lifetime's work.

14 The Universal problem and Duck Soup

IS THE EARTH TOO OLD?

As we have seen, by the 1930s the geologists finally were reasonably happy, following much soul-searching and research, with the notion that the Earth was approximately 3,000 million years old. With this harmonious consensus, one might have imagined that the debate on the age of Earth would then have died down, and the whole topic would have been sent into retirement in some quiet backwater.

The geologists might have been pleased with themselves, but the astronomers were not, because they had decided that the Universe was 2,000 million years old. How could the Earth be 1,000 million years older than the Universe? This would certainly be impossible, and so the findings of the geologists must have been at fault: after all the astronomers had a long history of observations with which to reinforce their contentions.

Stephen Brush of the University of Maryland has dissected this topic in some detail, and notes that the scientists tackled the problems of the age of the Earth and Universe from four perspectives. They brought to bear atomic physics and the study of radioactivity as already noted; stellar astronomy in which research was being rapidly advanced through the development of larger and more powerful telescopes with which to scan the night sky; theoretical physics and cosmology, which is the science of the Universe as a whole; and planetary geology, the study of the planets of our Solar System.

By the 1920s astronomy was gripped by a new theory that the Universe was expanding in size. It was recognised that distant galaxies were rapidly moving faster than those closer to us, and this was demonstrated in the distant galaxies where the degree of redshift was greater. A redshift is where the light emitted from the galaxies is moved to the longer-wavelength (redder) end of the light spectrum. Edwin Powell Hubble (1889–1953), now immortalised by a telescope bearing his name, studied these redshifts and so produced the evidence that led to the theory that the Universe was expanding. Hubble together with Milton La Salle Humason (1891–1971) calculated the time that it would take to pull all the Universe back to its starting point from where it expanded, and found that this would have taken 1,800 million years, which by inference was the age of the Universe. Computed distances of galaxies became a convenient measure of time. Hubble and Humason's date was in stark contrast to the timescale suggested by Sir James Hopwood Jeans (1877–1946) who suggested that the stars were of enormous age: 70,000,000,000,000 years!

The anomaly between the age based on the expanding Universe and the age of the Earth was first seriously raised three years before his death by Willem de Sitter (1872–1934), the Professor of Astronomy at Leiden, at a British Association for the Advancement of Science meeting. He pessimistically predicted that 'I do not think it will ever be found possible to reconcile the two time scales'. Two years later he argued that ideas on stellar evolution needed some revision, and cleverly postulated that if the Earth had formed through the collision of two stars, a concept then in vogue, that the chances of this happening were far greater when the Universe was small, at a time just before it had begun to expand. Therefore the age of the Universe and the Earth were the same, and this assertion was plausible according to de Sitter given the order of error in the geologists' chronological calculations. Jeans and his collaborators noted in 1935 that both terrestrial and astronomical data coalesced on a date of 2,100 million years. So all seemed to be quite satisfactory.

But by 1947 the accepted lower limit for the age of the Earth, as determined by the geologists using lead isotope data, and cited by Holmes, was 3,350 million years, and the difference between this figure and Hubble's 1,800 million years had increased to an unacceptable size. Were the geologists creating trouble again? In the following decade and a half the astronomers and physicists continued their investigations using more powerful telescopes which allowed them to see further into space than ever before. By the early 1950s they realised that the Universe was bigger than hitherto imagined, and that distance could be converted into time: at least 10,000 million years. Now in 1950 the chronological sequence made sense: Universe (10,000 million years old) → Earth (3,350 million years old), and all factions, geologists, physicists and astronomers could re-establish an amicable relationship.

By 1958 Allan Rex Sandage (b. 1926) had added another 3,000 million years to the astronomers' figure, and was touting an age for the Universe of 13,000 million years, which he based on research into the spectral characteristics of stellar clusters. Later he went on to suggest that the Universe expands with a 'Big Bang' and contracts with a 'Big Crunch' over and over again every 82 billion years or so. The idea later went out of fashion, but his estimate of 13,000 million years as the actual age of the Universe has so far stood the test of four decades of time.

THE PREPARATION OF DUCK SOUP

Sometime early in 1951 the geochemist Harrison Scott Brown (1917–1986) went looking for one of his new graduates at the Institute of Nuclear Studies at the University of Chicago so he could discuss his research topic with him. Brown was interested in the new field of nuclear geochemistry and was part of a group of superb students and academics, some of whom had worked on the Manhattan Project that produced the first atomic bomb. This group tackled problems such as using carbon-14 for dating, a technique developed by Willard Frank Libby (1908–1980) that has limited geological application but yielded a precise archaeological chronology. Carbon-14 is only suitable for material less than 50,000 years old because its half-life is so short. For this work Libby was awarded the Nobel Prize in Chemistry in 1960. Before arriving in Chicago, Harold Clayton Urey (1893–1981) had discovered deuterium for which he received the 1934 Nobel Prize in Chemistry, and he had also derived many dates for

Figure 14.1   Clair Cameron Patterson (1922–1995). By kind permission of Laurie Patterson.

geological materials from helium. In the Windy City he then pioneered the use of isotopes for the estimation of past temperatures: this scheme is still of vital importance as scientists try to understand the dynamics of global warming. Several others including Gerard Wasserburg, now of the California Institute of Technology (Caltech), worked on potassium/argon dating.

The beneficiary of Brown's visit was Clair Cameron Patterson (1922–1995) (Figure 14.1) who perhaps not surprisingly was known as 'Pat' to his friends. Born outside Des Moines, Iowa, a town that spawned the travel writer Bill Bryson, Patterson was first introduced to science when his mother gave him a rudimentary chemistry set. Captivated, he had taken the first step on the path that led him to membership of the National Academy of Sciences. He and his future wife Laurie studied at the University of Iowa, after which they both moved to Chicago and the Manhattan Project for which they were dispatched off to work in Tennessee for a period. Following the cessation of global hostilities he began a doctoral study at Chicago under the supervision of Brown. It was at about this time that Brown proposed the ground-breaking research. Patterson recalled just before his death that his supervisor breezed into his room and started talking about the direction in which he wanted the research to proceed. Brown set Patterson and another student George Tilton the task of working out how to determine the age of the mineral zircon by working out the proportions of minute amounts of uranium and its decay product lead found within the tiny zircon crystals. Tilton was set to work on the uranium while Patterson's role was to tackle lead, and he had the difficult task of measuring its different isotopic compositions. Brown acknowledged the difficulty of this work, and dangled a carrot in front of Patterson. If Patterson could complete the task, then he would be asked to determine the composition of lead isotopes in an iron meteorite. Brown had concluded that meteorites were coeval with the Earth and therefore that meteorite dating would yield the age of the Earth. If this could be achieved the student would become famous. Naturally Patterson agreed to this exciting challenge, to which Brown responded: 'It will be duck soup, Patterson'. It was just as well that it was Patterson and not the Marx Brothers who would be entrusted with this work.

It was a few years, however, before Patterson was in a position to test Brown's ideas relating to the lead contained in meteorites. He had to complete his Ph.D. first, and in 1951 completed his thesis on the composition of lead isotopes contained in a granite that proved to be 1,000 million years old. He then embarked on the one-year research programme based on meteorites that was to copper-fasten his scientific reputation.

TAKING THE LEAD OUT AND THE MAGIC NUMBER
As was known by 1950, the Earth contained various lead isotopes, some radiogenic, formed by the decay of uranium, and one stable non-radiogenic. Some lead was 'primeval' as shown by Al Nier and his colleagues, and had been present at the formation of the Earth, but researchers were unsure of the original isotopic composition of this ancient lead, and so were unable to compute ages accurately based on

it. Brown realised that the answer to this problem might lie in iron meteorites which he knew contained lead but did not contain any uranium, and so the lead must have been present when the Solar System and the Earth first formed. If the research team could determine the isotopic composition of meteoritic lead they could use this information to plug into the Holmes–Houtermans Model and derive the true age of the Earth.

Patterson began his post-doctoral fellowship at Chicago by submitting a proposal to the Atomic Energy Commission for funding which would allow him to carry out this work. The proposal was rejected, but Brown was having none of it: he resubmitted the proposal in his own name and Patterson received the funds. Within a year Brown was appointed to a position at the California Institute of Technology and so Patterson and his family migrated to the West Coast. He was to remain in the employ of Caltech for the remainder of his career.

Brown built new facilities in Pasadena and installed Patterson where he succeeded in isolating the primeval lead from the Canyon Diablo meteorite (Figure 14.2). This is probably the most famous meteorite to fall in North America: it crashed into the Earth some 50,000 years ago and formed the impressive crater in Arizona. On impact huge volumes of dust would have been thrown sky high, in a manner similar to that of a nuclear explosion, and fragments of the rock and meteorite fell back to the ground. Bits of the meteorite were thrown several miles into an adjacent canyon from which the meteorite derived its name. To extract the lead Patterson had to slice the meteorite open, and pick out some of the black sulphide that occurred in small pockets. He then dissolved this in acids and picked out the tiny fragments of lead which he placed securely in a small glass vial. Carefully packing up his samples of isolated lead, Patterson flew to Chicago in early 1953 where he stuck it into a mass spectrometer in Mark Inghram's laboratory at the Argonne National Laboratory run by the University of Chicago. Inghram (1919–2003) invented and modified these instruments, which according to Jerry Wasserburg 'were his favorite shovels for excavating new areas'. Making sure that

Figure 14.2   Fragment of the Canyon Diablo meteorite from Arizona (top) and two pieces of the Henbury meteorite from Northern Territories, Australia (bottom). Clair Patterson reported on the age of these meteorites in his 1956 paper. The Arizona sample is 50 mm wide (Geological Museum, Trinity College Dublin).

everything was in place, Patterson set the machine running and at the press of some buttons out came a printout that gave the isotopic composition of the meteoritic lead. Taking this new information Patterson grabbed a pencil and scribbled down some calculations. The figure on his piece of paper read: 4.5 billion years. Could this be right? He looked again. Yes! The calculations were correct: 4.5 billion years, or put another way, 4,500 million years. He had calculated the actual age of the Earth. Patterson was so excited by this that when he arrived at his parents' house shortly afterwards for a flying visit, he had to get his mother to admit him to a local hospital as he thought he was having a heart attack.

In late September 1953 Patterson addressed an audience attending the Conference on Nuclear Processes in Geologic Settings that had convened at Williams Bay, Wisconsin, and announced his results: the two ages he had derived from the Canyon Diablo meteorite were 4,510 and 4,570 million years. He presented the results again at the annual meeting of the Geological Society of America held that year on 9 to 11 November in Toronto, and attended by several thousand geologists who were left in no doubt that the Earth was greater than 4 billion years old. A scientific reporter picked up on this, and on 2 February 1954 the *New York Times* carried an article under the banner 'Atom study gives Earth age of 4.5 billion years'. Now both the scientific and general public knew about this remarkable piece of geochronological detective work.

However, the age of the Earth was based on samples taken from one meteorite only. Would others reveal concordant ages? Following a suggestion by his colleague Leon Silver, Patterson assembled a suite of different meteorites (Table 14.1). Two of them, the Forest City

Table 14.1 *Meteorites analysed by Clair Patterson whose ages were reported in his 1953*[a] *and 1956 papers.*

Canyon Diablo meteorite, Barringer Crater, Coconino County, Arizona. Found 1891. Iron.[a]

Henbury meteorite, Northern Territories, Australia. Fell 1931 [>2000 kg collected]. Iron.

Forest City meteorite, Winnebago County, Iowa. Fell 2 May 1890 [c. 122 kg]. Stone (Chondrite).[a]

Modoc meteorite, Scott County, Kansas. Witnessed fall 2 September 1902. Stone (Chondrite).[a]

Nuevo Laredo meteorite, Nuevo Laredo, Tamaulipas State, Mexico. Fell 1930. Stone (Achondrite).[a]

Age of meteorites and the earth

CLAIRE PATTERSON
Division of Geological Sciences
California Institute of Technology, Pasadena, California

(*Received* 23 *January* 1956)

Abstract—Within experimental error, meteorites have one age as determined by three independent radiometric methods. The most accurate method (Pb^{207}/Pb^{206}) gives an age of $4·55 \pm 0·07 \times 10^9$ yr. Using certain assumptions which are apparently justified, one can define the isotopic evolution of lead for any meteoritic body. It is found that earth lead meets the requirements of this definition. It is therefore believed that the age for the earth is the same as for meteorites. This is the time since the earth attained its present mass.

Figure 14.3   Title and abstract of Clair Patterson's 1956 paper on meteorites and the age of the Earth published in *Geochima et Cosmochimica Acta* **10** (1956) 230–237. This paper gained widespread circulation and the findings it reported were rapidly accepted by the scientific community. In this paper Patterson's first name is spelt 'Claire', but throughout his later life he dropped the 'e'.

meteorite from his home state of Iowa and the Modoc meteorite from Kansas, were stony meteorites which contained small glassy spheres called chrondrules, while another stony meteorite, the Nuevo Laredo meteorite from Mexico, was selected because it lacked chondrules (achrondrite). All produced ages close to 4,500 million years: Forest City 4.500; Nuevo Laredo 4,600; and Modoc 4,400. In 1956 Patterson published his classic paper (Figure 14.3) that reported on these ages and added another derived from the Henbury meteorite of Australia (Figure 14.2). He plotted the lead ratios for these meteorites and they formed a line or isochron whose slope gave an age of 4,550 million years. He noted that his uranium/lead dates were confirmed by colleagues including John Reynolds and Jerry Wasserburg a year earlier using the potassium/argon method. Crucially, he added the lead ratio for a modern marine sediment and this also plotted on the isochron, thus proving that the lead in meteorites and terrestrial material were one and the same age. Consigned to history were the estimates of 3,000 million years based on terrestrial lead.

The Earth is 4,550 million years old.

THE SHOULDERS OF GIANTS

Later in his career Patterson made the study of lead contamination of the environment his speciality, and we owe him a great debt for doing so much to highlight the dangerous health implications posed by his element. In 1992 he retired from the faculty of the California Institute of Technology and this occasion was marked with a two-day conference on 'Topics in Global Geochemistry' held on 3 and 4 December in Pasadena, California, and attended by fifty friends and scientists. Arising out of this meeting the journal *Geochimica et Cosmochimica Acta* published a special volume dedicated to him in 1994. The publication of a series of papers from the conference in this journal was only right, given it was in its pages that Patterson's classic paper on meteorites and the age of the Earth had appeared thirty-eight years earlier.

Patterson contributed three papers to the conference proceedings: one jointly authored piece examined the sources of lead in the surface waters of the North Atlantic; another, the final paper in the compilation, discussed the separation of thought processes in the human brain; while the last was an examination of the worth of scientific knowledge. He perceived a loss of integrity in modern geological research as he lamented that latter-day geologists were concerned with the acquisition of large data sets and models which provided solutions to open and closed cases. These researchers, he argued, failed to grasp the sense of continuum in their science.

We have all heard the expression first used by Isaac Newton, that he had 'stood on the shoulders of giants' – the contributions of earlier scientists and thinking men and women who provided the platform on which he could advance scientific understanding. Patterson in this volume articulated the same sentiments. He humbly acknowledged that he had become a scientist when he had been fortunate to be the first person to experience the numbing realisation of knowing the true age of the Earth. But at the precise moment of making this discovery he was unable to accept that **he** had done it. He could not jump for joy and shout, 'Look at what I have done!', but instead, he discerned 'a strong subconscious sense of obligation to the generations-old community of scientific minds for bequeathing this gift of glorious emotion to the discoverer from their treasury of scientific knowledge built over centuries.'

For centuries humankind had been attempting to discover the age of the Earth, and in 1953 Clair Patterson gave us the answer, but in his answer is their answer.

This is their story.

Bibliography

General works consulted

The following are useful introductions to the history of geology, while some are more focused on the history of research on the age of the Earth.

Adams, Frank Dawson *The Birth and Development of the Geological Sciences* (New York: Dover Publications, 1954)

Albritton, Claude C. Jr, *The Abyss of Time: Changing Conceptions of the Earth's Antiquity after the Sixteenth Century* (Los Angeles: Jeremy P. Tarcher, 1986)

Brush, Stephen G. *Transmuted Past: The Age of the Earth and the Evolution of the Elements from Lyell to Patterson* (New York: Cambridge University Press, 1996)

Dalrymple, G. Brent *The Age of the Earth* (Stanford University Press, 1991)

Ellenberger, François *History of Geology 1: From Ancient Times to the First Half of the XVII Century* (Rotterdam: A. A. Balkema, 1996)

Ellenberger, François *History of Geology 2: The Great Awakening and its First Fruits – 1660–1810* (Rotterdam: A. A. Balkema, 1999)

Geikie, Archibald *The Founders of Geology* (New York: Dover Publications, 1962)

Gillispie, Charles C. *Genesis and Geology* (Cambridge MA: Harvard University Press, 1996)

Good, Gregory A. (ed.) *Sciences of the Earth: an Encyclopedia of Events, People, and Phenomena* (2 vols) (New York & London: Garland Publishing, 1998)

Haber, Francis C. *The Age of the World: Moses to Darwin* (Baltimore: Johns Hopkins Press, 1959)

Harper, C. T. (ed.) *Geochronology: Radiometric Dating of Rocks and Minerals* (Stroudsburg PA: Dowden, Hutchinson & Ross, Inc., 1973)

Holland, Charles Hepworth *The Idea of Time* (Chichester: J. Wiley & Sons, 1999)

Holmes, Arthur *The Age of the Earth* (London & New York: Harper & Brothers, 1913)

Hurley, Patrick *How Old is the Earth?* (London: Heinemann, 1960)

Lewis, Cherry L. E. & Knell, Simon J. (eds.), *The Age of the Earth: from 4004 BC to AD 2002* (London: Geological Society Special Publication 190, 2001)

Mather, Kirtley F. (ed.), *Source Book in Geology 1900–1950* (Cambridge MA: Harvard University Press, 1967)

Mather, Kirtley F. & Mason, Shirley L. (eds.), *A Source Book in Geology 1400–1900* (Cambridge MA: Harvard University Press, 1970)

Oldroyd, David R. *Thinking About the Earth: a History of Ideas in Geology* (Cambridge MA: Harvard University Press, 1996)

Schneer, Cecil J. (ed.) *Toward a History of Geology* (Cambridge MA & London: MIT Press, 1969)

Thompson, Susan J. *A Chronology of Geological Thinking from Antiquity to 1899* (Metuchen NJ & London: Scarecrow Press, 1988)

Ward, Dederick C. & Carozzi, Albert V. *Geology Emerging: a Catalog Illustrating the History of Geology (1500–1850) from a Collection in the Library of the University of Illinois at Urbana-Champaign* (Urbana-Champaign: University of Illinois Library, 1984)

York, Derek *In Search of Lost Time* (Bristol: Institute of Physics Publishing, 1997)

York, Derek & Farquhar, Ronald M. *The Earth's Age and Geochronology* (Oxford: Pergamon, 1972)

Young, Davis A. *Christianity and the Age of the Earth* (Thousand Oaks CA: Artisan Sales, 1988)

Papers on aspects of this story and others in the history of geology may be found in *Earth Sciences History, Journal of Geological Education, Isis, British Journal for the History of Science*, and *Annals of Science* among others. The *Oxford Dictionary of National Biography* (Oxford University Press, 2004) remains invaluable, containing biographical information on many clerics, thinkers, travellers, physicists, geologists, earth scientists and oddities. In equal measure C. C. Gillespie (ed.) *Dictionary of Scientific Biography*, 16 vols (New York: Charles Scribner & Sons, 1970–1976) and B. Lightman (ed.) *Dictionary of Nineteenth-Century British Scientists*, 4 vols (Bristol: Thoemmes Continuum, 2004) are indispensable sources of biographical information. The National Academy of Sciences in Washington maintains biographies of many of its members, a number of whom feature large in this story. For Irish scientists see C. Mollan, W. Davis and B. Finucane (eds.) *Irish Innovators in Science and Technology* (Dublin: Royal Irish Academy, 2002).

Chapter 1. The ancients: early chronologies

The quotation from Lucretius is taken from John Joly's essay 'The birth-time of the World', *Science Progress* **9** (1914), 37 (this was reprinted in 1915 in his book of the same name: *The Birth-time of the World and Other Scientific Essays* (London: T. Fisher Unwin, 1915) pp. xv, 307). A useful treatise on early cosmological thinking of various civilisations is that by Dick Teresi, *Lost Discoveries* (New York: Simon &

Schuster, 2003), pp. ix, 453. In the *New Larousse Encyclopedia of Mythology*, 12th impression (London: Hamlyn, 1977), pp. xi, 500, R. Graves gives accounts of early ideas from fourteen civilisations, including Egypt, Babylonia, Greece, Germany and Scandinavia, India, the Far East, Africa and Oceania. F. A. Wright's *Lemprière's Classical Dictionary* (London: Routledge & Kegan Paul, 1951) is an invaluable source of information about the classical world. Manfred Lurker's excellent volume *The Gods and Symbols of Ancient Egypt* (London: Thames & Hudson, 1974), pp. 142 documents, explains and illustrates the significance of many deities and customs. Dennis Dean's essay 'The Age of the Earth controversy: beginnings to Hutton', *Annals of Science* **38** (1981), 435–456 gives a brief but useful synopsis of pre-Christian thinking about the Earth's age.

Chapter 2. Biblical calculations

The most recent biographical study of Theophilus is that by Rick Rogers, *Theophilus of Antioch: The Life and Thought of a Second-Century Bishop* (Lanham MD: Lexington Books, 2000).

Useful accounts of the work of James Ussher and his life are given by J. A. Carr, *The Life and Times of Archbishop Ussher* (London: Gardner, Darton & Co., 1895); Robert Buick Knox, *James Ussher, Archbishop of Armagh* (Cardiff: University of Wales Press, 1967); and Philip Styles, 'James Ussher and his times', *Hermathena* **88** (1956), 12–33. Ussher's works were reprinted in 17 volumes by C. R. Elrington, *The Whole Works of the Most Rev. James Ussher: With a Life of the Author, and an Account of His Writings* (Dublin: Hodges & Smith; London: Whittaker & Co., 1847–1864). For a discussion on the manuscripts of James Ussher held in the Library of Trinity College Dublin see Bernard Meehan's essay in Peter Fox (ed.), *Treasures of the Library, Trinity College Dublin* (Dublin: Royal Irish Academy, 1986), pp. 97–110.

Papers that deal with chronological work by Ussher, Lightfoot and others include R. L. Reese, S. M. Everett and E. D. Craun, 'The chronology of Archbishop James Ussher', *Sky and Telescope* **62**, part 5 (1981), 404–405; William R. Brice, 'Bishop Ussher, John Lightfoot and the age of creation', *Journal of Geological Education* **30**, part 1 (1982), 18–24; James Barr, 'Why the world was created in 4004 B.C.: Archbishop Ussher and biblical chronology', *John Rylands University Library of Manchester Bulletin* **67** (1985), 575–608; Stephen Jay Gould, 'Fall in the House of Ussher', *Natural History* **11** (1991), 12–21 [reprinted as Chapter 12, pp. 181–193 in *Eight Little Piggies: Reflections in Natural History* (London: Penguin Books, 1993)]; D. P. McCarthy, 'The biblical chronology of James Ussher', *Irish Astronomical Journal* **24**, part 1 (1997), 73–82, which recounts the

Irish Times proclamation of 1996; James Barr, 'Pre-scientific chronology: the Bible and the origin of the World', *Proceedings of the American Philosophical Society* **143**, part 3 (1999), 379–387; J. G. C. M. Fuller, 'Before the hills in order stood: the beginning of geology of time in England', in Lewis and Knell, *The Age of the Earth* (2001), pp. 15–23; and J. G. C. M. Fuller, 'A date to remember: 4004 BC', *Earth Sciences History* **24**, part 1 (2005), 5–14.

A number of original seventeenth-century chronologies were consulted (with profit), including:

Thomas Allen, *A Chain of Scripture Chronology; from the Creation of the World to the Death of Jesus Christ. In VII. Periods* (London: Printed by Thomas Roycroft for Francis Tyton and for Nathanial Ekins, 1659)

R[oger] D[rake], *Sacred Chronologie, drawn by Scripture Evidence al-along that vast body of Time, (containing the space of almost four thousand Years) From the Creation of the World, to the passion of our Blessed Saviour* (London: Printed by James and Joseph Maxex for Stephen Bowtell, 1648)

John Lightfoot, *A Few, and New Observations, upon the Booke of Genesis. the most of them certaine, the rest probable, all harmelesse, strange, and rarely heard off before* (London: Printed by T. Badger, 1642)

John Lightfoot, *The Harmony of the Foure Evangelists; among themselves, and with the Old Testament* (London: Printed by R. Cotes from Andrew Crooke, 1644)

William Nisbit, *A Scripture Chronology, wherein the principall Periods of Time from the Creation of the World to the death of Christ, are included, and many questions of great importance resolved* (Printed for Joshua Kirton, 1655)

James Ussher, *Annales veteris testamenti, a prima mundi origine deducti: una cum rerum Asiaticarum et Aegyptiacarum chronico, a temporis historici principio usque ad Maccabaicorum initia producto* (London: Printed by J. Flesher for L. Sadler, G. Bedell and J. Crook & J. Baker, 1650)

James Ussher, *The annals of the world deduced from the origin of time, and continued to the beginning of the Emperour Vespasians reign, and the totall Destruction and Abolition of the temple and common-wealth of the Jews: containing the historie of the Old and New Testament, with that of the Macchabees, also the most memorable affairs of Asia and Egypt, and the rise of the empire of the Roman Caesars under C. Julius, and Octavianus: collected from all history, as well sacred, as prophane, and methodically digested* (London: Printed by E. Tyler, for J. Crook, and for G. Bedell, 1658).

Authoritative accounts of the origin of the Julian Period and the mathematics on which it was based is given by R. L. Reese, S. M. Everett and E. D. Craun, 'The origin of the Julian Period: an application of congruences and the Chinese Remainder

Theorem', *American Journal of Physics* **49**, part 7 (1991), 658–661. A similar treatment is found in R. L. Reese and S. M. Everett, 'J. J. Scaliger and the Julian Period', *Griffith Observer* (May 1981), 17–19.

Chapter 3. Models of Aristotelian infinity and sacred theories of the Earth

Descartes is the subject of many books and monographs. Stephen Gaukroger, John A. Schuster and John Sutton's *Descartes' Natural Philosophy* (London: Routledge, 2000) and Daniel Garber's *Descartes Embodied: Reading Cartesian Philosophy Through Cartesian Science* (Cambridge University Press, 2001) provide a great deal of information about this philosopher and his thinking, and may be read with profit. Joscelyn Godwin's *Athanasius Kircher: a Renaissance Man and the Quest for Lost Knowledge* (London: Thames & Hudson, 1979) is the most up-to-date treatment of this versatile man and priest. The account of his life and work in James J. Walsh, *Catholic Churchmen in Science* (New York: Books for Libraries, 1968, reprint of 1908 original edition) is moderately useful, although it fails to document or discuss his work in geology. A synopsis of the Cartesian and Kircherian models of the Earth appears in Ezio Vaccari, 'European views on terrestrial chronology from Descartes to the mid-eighteenth century', Lewis and Knell, *The Age of the Earth* (2001), pp. 25–37. Vaccari also discusses the reception in Europe of the ideas presented in various seventeenth century books by the biblical theorists. Oldroyd, *Thinking About the Earth* (1996) gives a good outline of Kircher's geological ideas.

Excellent accounts of the sacred theories of Burnet, Woodward and Whiston are given in Gordon L. [Herries] Davies, *The Earth in Decay* (New York: Elsevier, 1969); Roy Porter, *The Making of Geology: Earth Science in Britain 1660–1815* (Cambridge University Press, 1977); and Rhoda Rappaport, *When Geologists Were Historians, 1665–1750* (Ithaca: Cornell University Press, 1997). Stephen Jay Gould, *Time's Arrow, Time's Cycle: Myth and Metaphor in the Discovery of Geological Time* (Cambridge MA: Harvard University Press, 1997) discusses Burnet's *Telluris* in some detail in Chapter 2. The quotations from Thomas Burnet are taken from Dennis R. Dean, *Annals of Science* (1981), and that of the Bishop of Hereford from Davies *The Earth in Decay* (1969), p. 73. Burnet's *The Sacred Theory of the Earth* was reprinted in 1965 with a commentary by Basil Willey (London: Centaur Press, 1965). Two recent books on Whiston and his thinking are those by James E. Force, *William Whiston, Honest Newtonian* (Cambridge University Press, 1985) and Maurice Wiles, *Archetypal Heresy: Arianism Through the Centuries* (Oxford: Clarendon Press, 1996). Joseph M. Levine's *Dr Woodward's Shield: History,*

Science, and Satire in Augustan England (Berkeley: University of California Press, 1977) provides a good account of the social and scientific circles in which Woodward lived and worked. A facsimile edition of his *Brief instructions for making observations in all parts of the World* (1696) was published by the Society for the Bibliography of Natural History (now the Society for the History of Natural History) in 1973 and complemented with an introduction by Victor Eyles. John Woodward's museum complete with his original cabinets and specimens is described by David Price in the *Journal of the History of Collections* **1**, part 1 (1989), 79–95. The provenance of some of the minerals acquired by Woodward from the Carpathians is discussed by Miklós Kázmér in the same journal, vol. **10**, part 2 (1998), 159–168. Woodward's ideas on the nature of fossils are discussed by Martin Rudwick in *The Meaning of Fossils: Episodes in the History of Palaeontology*, 2nd edition (University of Chicago Press, 1985).

Chapter 4. Falling stones, salty oceans, and evaporating waters: early empirical measurements of the age of the Earth

The authoritative source on Edward Lhwyd is that by R. T. Gunther, *Early Science in Oxford. Life and Letters of Edward Lhuyd* (Oxford: privately published by the author, 1945). In it Gunther reproduces the fine biographical memoir by Richard Ellis that first appeared in the *Transactions of the Cymmrodorion Society* for 1907. Equally the paper by Brynley F. Roberts, 'In search of Edward Lhwyd', *Archives of Natural History* **16**, part 1 (1989), 49–57, is useful. Lhwyd's masterful illustrated catalogue of fossils *Lithophylacii Britannici Ichnographia* and the difficulties he experienced getting it published are discussed in some detail by Marcus Hellyer in *Archives of Natural History* **23**, part 2 (1996), 43–60, while the various issues of this work are described by M. E. Jahn in the *Journal of the Society for the History of Natural History* **6** (1972), 86–97. Details of Lhwyd's Irish botanical observations are given by M. E. Mitchell, 'Irish botany in the seventeenth century', *Proceedings of the Royal Irish Academy* **75B** (1975), 275–284. Lhwyd's role in the description of early dinosaur bones in Britain is recounted by Justin B. Delair and William A. S. Sarjeant, 'The earliest discoveries of dinosaur bones: the records re-examined', *Proceedings of the Geologists' Association* **113**, part 3 (2002), 185–197. They also reproduce a portrait of Lhwyd from a document in Merton College, Oxford; a similar but clearer image reproduced here is given in Robert M. Owens, 'Trilobites in Wales', *National Museum of Wales, Geological Series* **7** (1984), 1–22. Discussion of Brookes' name *Scrotum humanum* is contained in L. B. Halstead and W. A. S. Sarjeant, '*Scrotum humanum* Brookes–the earliest name for a dinosaur?', *Modern Geology* **18** (1993), 221–224. For additional discussion of Lhwyd's thoughts on

Llanberis, I am grateful to Michael Roberts whose comments on the episode I read online.

The most comprehensive treatment of Edmond Halley and his life and work is that by Alan Cook, *Edmond Halley: Charting the Heavens and the Seas* (Oxford: Clarendon Press, 1998). Halley's paper outlining his interesting scheme on the increasing salinity of closed lakes was published in the *Philosophical Transactions of the Royal Society* **29**, number 344 (1715), 296–300. Halley's paper and its ideas were forgotten until 1910 when they were brought to light by the American geochronologer G. F. Becker, 'Halley on the age of the ocean', *Science* **31** (1910), 459–461.

For Maillet see Chapter 6 in Albritton's wonderful book *The Abyss of Time* (1986); the earlier volume by Francis C. Haber, *The Age of the World: Moses to Darwin* (Baltimore: John Hopkins University Press, 1959), and Ezio Vaccari, 'European views on terrestrial chronology from Descartes to the mid-eighteenth century', in Lewis and Knell, *The Age of the Earth* (2001), pp. 25–37. A modern edition of de Maillet's *Telliamed* edited by Albert V. Carozzi (Urbana: University of Illinois Press, 1968) is closer to the manuscript version in that the alterations incorporated by le Mascrier in the first published edition of 1748 have been highlighted.

Chapter 5. Thinking in layers: early ideas in stratigraphy

The earliest English translation of Nicolaus Steno's *Prodromus* appeared in 1671, but that by John Garrett Winter published in 1916 by The Macmillan Company, New York remains second to none – this volume also contains a useful biography of Steno and various explanatory notes on his text. For two good biographies of Steno see: Harald Moe *Nicolaus Steno: An Illustrated Biography. His Tireless Pursuit of Knowledge, His Genius, His Quest for the Absolute* (Copenhagen: Rhodos, 1994), and Erik Kennet Pålsson *Niels Stensen: Scientist and Saint* (Dublin: Veritas, 1988). A recent biography aimed at the popular market is that by Alan Cutler *The Seashell on the Mountaintop: A Story of Science, Sainthood, and the Humble Genius Who Discovered a New History of the Earth* (London: Heinemann, 2003). The quotation of the lines on the plaque at Steno's tomb is taken from the translation given in James J. Walsh, *Catholic Churchmen in Science* (New York: Books for Libraries Press, 1968, a reprint of the 1908 edition), p. 160–161.

For some analysis of Steno's geological work see Gordon L. Herries Davies, 'The Stenoian Revolution', pp. 44–49, in Gaetano Giglia, Carlo Maccagni and Nicoletta Morello (eds.) *Rocks, Fossils and History* (Firenze: Festina Lente, 1995); Charles Hepworth Holland, *The Idea of Time* (Chichester: J. Wiley & Sons,

1999); Gary D. Rosenberg, 'An artistic perspective on the continuity of space and the origin of modern geologic thought', *Earth Sciences History* **20**, part 2 (2001), 127–155.

An account of the second International Geological Congress is given by Gian Battista Vai, 'The second International Geological Congress, Bologna, 1881', *Episodes* **4**, part 1 (2004), 13–20.

In recent years four biographies of Robert Hooke have appeared on the book-stands: Stephen Inwood, *The Man Who Knew Too Much: The Strange and Inventive Life of Robert Hooke 1635–1703* (London: Macmillan, 2002); Lisa Jardine, *The Curious Life of Robert Hooke: The Man Who Measured London* (London: Harper Collins, 2003); Jim Bennett, Michael Cooper, Michael Hunter and Lisa Jardine, *London's Leonardo: The Life and Work of Robert Hooke* (Oxford University Press, 2003); and Michael Cooper, *'A More Beautiful City': Robert Hooke and the Rebuilding of London After the Great Fire* (Stroud: Sutton Publishing, 2005).

The most comprehensive account of Hooke's geological work is contained in Ellen Tan Drake, *Restless Genius: Robert Hooke and his Earthly Thoughts* (Oxford University Press, 1996). She compared Hooke's work with that of Steno in an earlier paper: Ellen T. Drake and Paul D. Komar, 'A comparison of the geological contributions of Nicolaus Steno and Robert Hooke', *Journal of Geological Education* **29** (1981), 127–132 (from which I have taken Hooke's quotation on fossils); and recently she has discussed the influence of his birthplace on his geological ideas: E. T. Drake, 'The geological observations of Robert Hooke (1635–1703) on the Isle of Wight', in Patrick N. Wyse Jackson (ed.), *Geological Travellers: On Foot, Bicycle, Sledge or Camel, the Search for Geological Knowledge* (Staten Island, New York: Pober Publications, 2006). Other papers on Hooke's geological thinking include Gordon L. Davies, 'Robert Hooke and his conception of Earth history', *Proceedings of the Geologists' Association* **75**, part 4 (1964), 493–498; and David R. Oldroyd, 'Robert Hooke's methodology of science as exemplified in his Discourse on Earthquakes', *British Journal for the History of Science* **6** (1972), 109–130.

The Michell quote is taken from Archibald Geikie, *The Founders of Geology* (New York: Dover, 1962, reprint of the 1905 2nd edition); it is also reproduced in Roy Porter, *The Making of Geology: Earth Science in Britain 1660–1815* (Cambridge University Press, 1977). A synopsis of the ideas of John Strachey regarding the development of strata is contained in Barry D. Webby, 'Some early ideas attributing easterly dipping strata to the rotation of the Earth', *Proceedings of the Geologists' Association* **80**, part 1 (1969), 91–97. Strachey's use of cross-sections is discussed in J. G. C. M. Fuller, 'The invention and first use of stratigraphic cross-sections by John Strachey, F.R.S. (1671–1743)', *Archives of Natural History* **19**, part 1 (1992), 69–90.

The best biography of Giovanni Arduino available is the comprehensive volume by Ezio Vaccari: *Giovanni Arduino (1714–1795): il contributo di uno scienziato veneto al dibattito settcentesco sulle scienze della Terra* (Firenze: Leo S. Olschki, 1993). A translation into English of this scholarly work is urgently needed. For a discussion of Bergman's ideas on stratigraphy, see the paper by H. D. Hedberg, 'The influence of Torbern Bergman (1735–1784) on stratigraphy: a resumé', in Schneer, *Toward a History of Geology* (1969), pp. 186–191. In the same volume, pp. 242–256, Alexander M. Ospovat gives a detailed exegesis of Werner's stratigraphical classification in 'Reflections on A. G. Werner's "Kurze Klassifikation"'. Mott T. Greene, *Geology in the Nineteenth Century: Changing Views of a Changing World* (Ithaca & London: Cornell University Press, 1982) devotes much of one chapter to a discussion of Werner's ideas. Werner's quote on the endless span of geological time is from A. G. Werner, *Short Classification and Description of the Various Rocks. Translation of the German Text of 1786 by Alexander Ospovat* (New York: Hafner, 1971) and also reproduced in Claude C. Albritton Jr, 'Geologic Time', *Journal of Geological Education* **32** (1984), 29–37.

Chapter 6. An infinite and cyclical Earth and religious orthodoxy

A great deal has been written on James Hutton and his theory, starting with John Playfair's *Illustrations of the Huttonian Theory of the Earth*, published in Edinburgh in 1802. This edition was reprinted by Dover Books in 1956. Playfair also produced an early biographical account that was published in the *Transactions of the Royal Society of Edinburgh* in 1805. It too has been reissued, in George W. White (ed.) *Contributions to the History of Geology*, Vol. 5 (Connecticut: Hafner Publishing Company, 1970), and more recently, in 1997, by the RSE Scotland Foundation. The drawings sketched by John Clerk and others while on field work with Hutton were discovered in 1968 and reproduced in a fine slipcase that was accompanied by an explanatory book by Gordon Craig, Donald McIntyre and Charles Waterston, *James Hutton's Theory of the Earth: The Lost Drawings* (Edinburgh: Scottish Academic Press, 1978). Donald McIntyre has produced a large canon of work on Hutton that includes a comprehensive paper on his Edinburgh circle published in *Earth Sciences History* **16**, part 2 (1997), 100–157 (a *précis* of this paper was published in the volume that emanated from the James Hutton bicentenary conference held in Edinburgh: G. Y. Craig and J. H. Hull (eds.), *James Hutton – Present and Future* (London: Geological Society Special Publication 150, 1999)), and a booklet co-authored with Alan McKirdy, *James Hutton: The Founder of Modern Geology* (London: The Stationery Office, 1997). Interesting details of Hutton's Edinburgh house are given by Norman Butcher, 'James

Hutton's house at St John's Hill, Edinburgh', *Book of the Old Edinburgh Club*, new series, vol. 4 (1998 for 1997), 107–112, who also described the Hutton Memorial Garden in the *Newsletter of the History of Geology Group of the Geological Society of London*, number 16 (August 2002), 11–12. Gordon Herries Davies discusses Hutton and time in *The Earth in Decay* (1969), while David J. Leveson outlined Hutton's methodology in *Archives of Natural History* **23**, part 1 (1996), 61–77. However, if you wish to read only one volume on Hutton and his work, that by Dennis Dean would be beneficial: *James Hutton and the History of Geology* (Ithaca & London: Cornell University Press, 1992). Facsimiles of Hutton's 1785 abstract were published in White's *Contributions to the History of Geology* (1970), and in 1997 by the Edinburgh University Library. The former has an introduction authored by Victor A. Eyles, and the latter an accompanying introduction by Gordon Craig. The volume by White reproduces Hutton's 1788 paper in full; David Oldroyd has provided an annotated version of this seminal paper: 'Benchmark papers in the history of geology. 1. James Hutton's "Theory of the Earth" (1788)', *Episodes* **23** (2000), 196–200. Peter J. Wyllie has written several papers on the experimental work on the melting and recrystallisation of basalt carried out by Sir James Hall: 'Hutton and Hall on theory and experiments: the view after 2 centuries', *Episodes* **21**, part 1 (1998), 3–10; and 'Hot little crucibles are pressured to reveal and calibrate igneous processes', in Craig and Hull, *James Hutton – Present and Future* (1999), pp. 37–57. Although S. I. Tomkeieff's paper on the history of thought on unconformities is over forty years old it still has not been bettered: S. I. Tomkeieff, 'Unconformity – an historical study', *Proceedings of the Geologists' Association* **73**, part 4 (1962), 383–416. In this paper he quotes Hutton's 1795 descriptions of the discovery of the Arran unconformity, and some of that quotation is reproduced here.

The work of de Luc has been discussed by Martin Rudwick in 'Jean-André de Luc and nature's chronology', in Lewis and Knell *The Age of the Earth* (2001), pp. 51–60. An account of the Neptunist–Volcanist/Plutonist debate may be found in Gordon L. Herries Davies, 'The Neptunian and Plutonic Theories', in D. G. Smith (ed.), *Cambridge Encyclopaedia of Earth Sciences* (Cambridge University Press, 1981). William Richardson's views are elaborated in Patrick N. Wyse Jackson, 'Tumultuous times: geology in Ireland and the debate on the nature of basalt and other rocks of north-east Ireland between 1740 and 1816', in P. N. Wyse Jackson (ed.), *Science and Engineering in Ireland in 1798: A Time of Revolution* (Dublin: Royal Irish Academy, 2000), pp. 35–50. Martin Anglesea and J. Preston discuss and illustrate the work of Susanna Drury in '"A philosophical landscape": Susanna Drury and the Giant's Causeway', *Art History* **3** (1980), 252–273. The best discussion of the work of Desmarest remains that by Ken Taylor, 'Nicolas Desmarest and geology in the eighteenth century', in Schneer *Toward a History of Geology* (1969), pp. 339–356. G. H. Toulmin and his ideas have been treated by Gordon L. Davies, 'George Hoggart Toulmin and the Huttonian theory of the Earth', *Geological Society of America Bulletin* **78**, part 1 (1967), 121–123; by Roy S. Porter, 'Philosophy and politics of a geologist: G. H. Toulmin (1754–1817)', *Journal of the History of Ideas* **39**, part 3 (1978), 435–450; and by Dean in *James Hutton and the History of Geology* (1992), pp. 272–275.

A biographical account of the life and work of J. H. Balfour by R. D. Bellon appears in the *Dictionary of Nineteenth Century British Scientists*, vol. 1 (Bristol: Thoemmes Continuum, 2004), pp. 102–105.

Chapter 7. The cooling Earth

The most comprehensive biography of Buffon is that by Jacques Roger [translated by Sarah Lucille Bonnefoi], *Buffon: A Life in Natural History* (Ithaca & London: Cornell University Press, 1997) which first appeared in French as *Buffon, un philosophe au Jardin du Roi* (Paris: Librairie Arthème Fayard, 1989). Earlier biographies in English include Otis E. Fellows and Stephen F. Milliken, *Buffon* (New York: Twayne's World Authors Series 243, Twayne Publishers Inc., 1972). A good synopsis of his life is given in the *Dictionary of Scientific Biography*. Buffon's determinations of the age of the Earth are discussed by Lawrence Badash, 'The age-of-the-Earth debate', *Scientific American* **261** (1989), 90–96, and this and other aspects of his work are outlined in Chapter 7 of Albritton, *The Abyss of Time* (1986), pp. 78–88.

The quotation from Newton is taken from Dalrymple, *The Age of the Earth* (1991), p. 28. The tabulated heating and cooling times of Buffon's 'bullets' are embedded in the text found on pages 111 and 112 of volume 10 of the 1792 English edition of his work: *Barr's Buffon. Buffon's Natural History. Containing a Theory of the Earth, a general History of Man, of the Brute Creation, and of Vegetables, Minerals&c. From the French. With notes by the Translator. In ten volumes.* (London: J. S. Barr, 1792). Arthur Stinner in 'Calculating the age of the Earth and the Sun', *Physics Education* **37**, part 4 (2002), 296–305 checked Newton's cooling estimate of 50,000 using Stefan's law of radiation and derived a close match of 45,000 years.

An English translation of the part of *Époques* that describes in detail the characteristics of each of Buffon's seven epochs can be found in Mather and Mason *A Source Book in Geology* (1970) pp. 65–73. Useful information on Buffon's *Époques de la Nature* is given on a website produced by the Academy of Natural Sciences in Washington entitled 'Thomas Jefferson Fossil Collection' which also discusses Buffon's ideas on degeneracy of animal species and the reaction in

America to this theory (http://www.acnatsci.org/museum/jefferson/otherPages/degeneracy-01.html).

Jeff Loveland's essay on the three English translations of Buffon is given in 'Georges-Louis Leclerc de Buffon's *Histoire naturelle* in English, 1775–1815', *Archives of Natural History* **31**, part 2 (2004), 214–235.

A wonderful account of the long journey from Africa of the first giraffe to set foot in Paris was published as *Zarafa* by Michael Allin (London: Headline Book Publishing, 1998).

Chapter 8. Stratigraphical laws, uniformitarianism and the development of the geological column

The late John C. Thackray has provided a wonderful first-hand account of the early geological debates in the chambers of the Geological Society: 'To see the Fellows fight', *British Society for the History of Science Monographs* **12** (2003). The quote from Agassiz following the presentation of glacial observations in November 1840 is taken from a letter addressed to Sir Phillip Egerton reproduced on page 101.

An exhaustive list of the foundation dates of Geological Surveys is given in David R. Oldroyd, *Thinking About the Earth: A History of Ideas in Geology* (Cambridge MA: Harvard University Press, 1996), pp. 124–127.

For further information on William Smith see the account of his life by his nephew John Phillips. This was first published in 1844 but has recently been reprinted as *Memoirs of William Smith, LL.D.* (Bath Royal Literary and Scientific Institution, 2003) and contains an introduction to the life and times of the subject by Hugh Torrens together with a reprinting of his essay that first appeared as 'Timeless order: William Smith (1769–1839) and the search for raw materials 1800–1820', in Lewis and Knell, *The Age of the Earth* (2001), pp. 61–83. A short chronology of significant dates in Smith's life was compiled by Joan M. Eyles, *Proceedings of the Geological Society of London*, number 1657 (1969), 173–176. Recently small-scale facsimiles of his 1815 and 1820 maps have been made available by the British Geological Survey. Smith's cross-sections of 1819 were reproduced in 1995 in poster form by the American Association of Petroleum Geologists and the Geological Society of London together with an explanatory booklet by John G. C. M. Fuller, *"Strata Smith" and his Stratigraphic Cross Sections, 1819.*

Joe Burchfield's paper together with others on Charles Lyell appeared in D. J. Blundell and A. C. Scott (eds.), *Lyell: The Past is the Key to the Future* (Geological Society Special Publication 143, 1998). The most useful biographical treatment of Lyell's early life is that by Leonard Wilson, *Charles Lyell: The Years to 1841* (New Haven & London: Yale University Press, 1972); a now somewhat dated but nevertheless interesting biography is Edward B. Bailey's *Charles Lyell* (London: Thomas Nelson & Sons, 1962). For details of Lyell's travels in North America see Leonard Wilson, *Lyell in America: Transatlantic Geology, 1841–1853* (Baltimore: Johns Hopkins University Press, 1998), and two papers by Robert H. Dott Jr, 'Charles Lyell in America – his lectures, field work, and mutual influences, 1841–1853', *Earth Sciences History* **15**, part 2 (1996), 101–140, and 'Charles Lyell's debt to North America: his lectures and travels from 1841 to 1853', in Blundell and Scott *Lyell: The Past is the Key to the Future* (1998).

Christopher McGowan is a leading scholar of Mesozoic marine reptiles, and in his *The Dragon Seekers* (Cambridge MA: Perseus Publishing, 2001) tells the compelling story of their discovery in England by Mary Anning and others. Mary Anning was until recently a somewhat obscure character but she has been drawn out of the recesses of time by Hugh Torrens in his paper 'Mary Anning (1799–1847) of Lyme; "The greatest Fossilist the World ever knew"', *British Journal of the History of Science* **28** (1995), 257–284. She is also the subject of a wonderful children's book by Catherine Brighton, *The Fossil Girl* (London: Francis Lincoln, 1999).

The debate on the attributon of Scottish sandstones at Elgin is given in John A. Deimer, 'Old or New Red Sandstone? Evolution of a nineteenth century stratigraphic debate, northern Scotland', *Earth Sciences History* **15**, part 2 (1996), 151–166. The arguments between Geikie and Hicks are recounted by Paul Pearson and Chris Nicholas in 'Defining the base of the Cambrian: the Hicks–Geikie confrontation of April 1883', *Earth Sciences History* **11**, part 2 (1992), 70–80, and also in David R. Oldroyd, 'The Archean controversy in Britain: Part 1 – The rocks of St. David's', *Annals of Science* **48** (1991), 407–452.

W.B.N. Berry's *Growth of a Prehistoric Time Scale: Based on Organic Evolution* (Palo Alto & Oxford: Blackwell Scientific Publications, 1987) contains the best overview of the development of the geological column from a palaeontological perspective. A great deal has been written about the controversies relating to the delineation of some of the geological Periods. For lively and exhaustive accounts of the debates surrounding the Cambrian and Silurian see James Secord, *Controversy in Victorian Geology: the Cambrian–Silurian Dispute* (Princeton University Press, 1986), and for the Devonian arguments see Martin Rudwick, *The Great Devonian Controversy* (University of Chicago Press, 1985), which also briefly touches on Murchison's travels in Russia (pp. 376–379). The naming of the Pennsylvanian is discussed by William R. Brice, 'Henry Shaler Williams (1847–1918) and the Pennsylvanian Period', *Northeastern Geology and Environmental Sciences* **22**, part 4 (2000), 286–293.

Michael Collie and John Diemer have recently produced an edited account of Murchison's time spent in Russia: *Murchison's Wanderings in Russia: his*

Geological Exploration of Russia in Europe and the Ural Mountains, 1840 and 1841 (Keyworth: British Geological Survey, Occasional Publication 2, 2004).

For accounts of the competing Diluvialist, Fluvialism and Glacial theories see Davies, *The Earth in Decay* (1969), Charles C. Gillispie, *Genesis and Geology* (1996); and Dennis R. Dean, 'The rise and fall of the deluge', *Journal of Geological Education* **33** (1985), 84–93. The most valuable assessment of William Buckland's geological work is that by Nicolaas A. Rupke, *The Great Chain of History: William Buckland and the English School of Geology 1814–1849* (Oxford University Press, 1983).

An early account of the use of colour in geological maps is given by J. E. Portlock in his monumental *Geological Report on Londonderry, and Parts of Tyrone and Fermanagh* (Dublin: Andrew Milliken, 1843), pp. 8–12. David McMahon described the earliest geological map in 'The Turin Papyrus map: the oldest known map with geological significance', *Earth Sciences History* **11**, number 1 (1992), 9–12.

Chapter 9. 'Formed stones' and their subsequent role in biostratigraphy and evolutionary theory

Recent papers that discuss the earliest examples of life on Earth include those by J. W. Schopf, 'Microfossils of the Early Archean Apex chert: new evidence of the antiquity of life', *Science* **260** (1993), 640–646, and S. J. Mojzsis, G. Arrhenius, K. D. McKeegan, T. M. Harrison, A. P. Nutman and C. R. Friend, 'Evidence for life on Earth before 3,800 million years ago', *Nature* **384** (1996), 55–59. However, some of the conclusions regarding early life have been questioned: C. M. Fedo and M. J. Whitehouse, 'Metasomatic origin of quartz-pyroxene rock, Akilia, Greenland, and implications for Earth's earliest life', *Science* **296** (2002), 1448–1452; and M. D. Brasier, O. R. Green, A. P. Jephcoat *et al.* 'Questioning the evidence for Earth's oldest fossils', *Nature* **416** (2002), 28.

Useful treatments of the subject of fossils and folklore include M. G. Bassett, *'Formed Stones', Folklore and Fossils*, National Museum of Wales Geological Series **1** (1982) 1–32, and the papers by Kenneth P. Oakley, 'Folklore of fossils', *Antiquity* **39** (1975), 9–16; 117–125; M. E. Taylor and R. A. Robison, 'Trilobites in Utah folklore', *Brigham Young University Geology Studies* **23** (1976), 1–5; and G. Zammit Maempel, 'The folklore of Maltese fossils', *Papers in Mediterranean Social Studies* **1** (1989), 1–29. Paul D. Taylor of the Natural History Museum, London has established a webpage devoted to the subject (http://www.nhm.ac.uk/nature-online/earth/fossils/fossil-folklore/). The early use of fossils associated with burials has been documented by Patrick N. Wyse Jackson and Michael Connolly, 'Fossils as

Neolithic funereal adornments in County Kerry, southwest Ireland', *Geology Today* **18**, part 4 (2002), 139–143. Beliefs in classical times are recalled in Adrienne Mayor, *The First Fossil Hunters: Paleontology in Greek and Roman Times* (Princeton University Press, 2000) and in Nikos Solounias and Adrienne Mayor, 'Ancient references to the fossils from the land of Pythagoras', *Earth Sciences History* **23**, part 2 (2004), 283–296.

For a colourful account of the life and work of Alcide d'Orbigny see Philippe Taquet, *Un voyageur naturaliste Alcide d'Orbigny. Du nouveau monde ... au passé du monde* (Paris: Nathan and Muséum National d'Histoire Naturelle, 2002).

Cecil Schneer has published a webpage on William Smith from which various documents including his *Strata Identified by Fossils* may be downloaded: http://www.unh.edu/esci/wmsmith.html.

Samantha Weinberg gives a dramatic and highly readable account of the 1938 discovery of the coelacanth in *A Fish Caught in Time: The Search for the Coelacanth* (London: Fourth Estate, 1999).

An account of the ideas of Lamarck and Cuvier is given by Martin Rudwick, *The Meaning of Fossils; Episodes in the History of Palaeontology* (University of Chicago Press, 1985) from where (p. 119) I have taken the quotation regarding the work of the former. Rudwick also discusses early ideas in palaeontology and also those of Darwin on evolution. In *Georges Cuvier, Fossil Bones, and Geological Catastrophes* (University of Chicago Press, 1997) Rudwick provides translations and commentories on a number of Cuvier's more important publications.

Robert Chambers' *Vestiges of the Natural History of Creation* was republished in 1994 by Chicago University Press, and this edition contains an introduction by James Secord. A short but useful history of palaeontology is contained in Derek E. G. Briggs and Peter R. Crowther (eds.) *Palaeobiology: A Synthesis* (Oxford: Blackwell Scientific Publications, 1990).

Chapter 10. The hour-glass of accumulated or denuded sediments

A short account of Herodotus' geological observations are given in Frank L. Kessler, 'Sailing up the Nile River – in the company of Herodotus, World's first geologist/ geographer, 2500 years ago', *Houston Geological Society Bulletin* **47**, number 3 (2004), 14; 61.

John Phillips' ideas on the rates of sedimentation and their application to the Earth's chronology were first published in his presidential address to the Geological Society of London: 'The Anniversary Address of the President', *Proceedings of the Geological Society of London* in the *Quarterly Journal of the Geological Society* **16** (1860), xxvii–lv. The chronological part of the address was expanded and appeared as a book later that year: *Life on the Earth: Its Origin and Succession* (Cambridge & London: Macmillan, 1860). The leading authority on the geological work of John Phillips is Jack Morrell and a number of his publications have been recently republished in the collection *John Phillips and the Business of Victorian science* (Aldershot: Ashgate Publishing, 2005). Of most interest to this topic is his 2001 paper 'Genesis and geochronology: the case of John Phillip (1800–1874)', in Lewis and Knell, *The Age of the Earth* (2001), pp. 85–90.

A great deal has been written about Darwin's work in geology. Sandra Herbert, the foremost authority, has recently published a book on the topic: *Charles Darwin, Geologist* (Ithaca: Cornell University Press, 2005). However, Darwin's geological activities were first discussed by Archibald Geikie in his Rede Lecture of 1909; more recent papers, some of which also discuss his 1831 fieldwork, include those by Paul H. Barrett, 'The Sedgwick–Darwin geological tour of North Wales', *Proceedings of the American Philosophical Society* **118** (1974), 146–164; Sandra Herbert, 'Darwin as a geologist', *Scientific American* **254**, number 5 (1986), 94–101; Michael B. Roberts, 'I coloured a map: Darwin's attempts at geological mapping in 1831', *Archives of Natural History* **27**, number 1 (2000), 69–79; Peter Lucas, ' "A most glorious country": Charles Darwin and North Wales, especially his 1831 geological tour', *Archives of Natural History* **27**, number 1 (2002), 1–26; Sandra Herbert and Michael B. Roberts, 'Charles Darwin's notes on his 1831 geological map of Shrewsbury', *Archives of Natural History* **29**, number 1 (2002), 27–30; Sandra Herbert, 'Doing and knowing: Charles Darwin and other travellers', in Wyse Jackson, *Geological Travellers* (2006); and Paul N. Pearson and Christopher J. Nicholas, ' "Marks of Extreme Violence": Charles Darwin's geological observations on St Jago (São Tiago), Cape Verde Islands', in Wyse Jackson (2006), *ibid.*

Samuel Haughton's calculations were published in *Manual of Geology*, 3rd edn (London: Longman, Green, Reader & Dyer, 1871) and in S. Haughton, 'A geological proof that changes in climate in past times were not due to changes in position of the Pole; with an attempt to assign a minor limit to the duration of geological time', *Nature* **18** (1878), 266–268. The quotation from Arthur Holmes reworking Haughton's principle is from his paper 'The construction of a geological time-scale', *Transactions of the Geological Society of Glasgow* **21** (1947), 117–152.

The most influential papers originating from the United States on the subject of sediment accumulation and the age of the Earth were Charles Doolittle Walcott, 'Geologic time, as indicated by the sedimentary rocks of North America', *Journal of Geology* **1** (1893), 639–676; and Joseph Barrell, 'Rhythms and the measurements of geologic time', *Geological Society of America Bulletin* **28** (1917), 745–904. Many of the American contributions are discussed in detail by Ellis L. Yochelson and Cherry Lewis: 'The age of the Earth in the United States

(1891–1931): from the geological viewpoint', in Lewis and Knell, *The Age of the Earth* (2001), pp. 139–155, who also note that Joseph Barrell's paper was first read in two parts at the University of Illinois in January 1912, four years before it was reprised at the Albany meeting.

The standard biography of Walcott is the two-volume set by Ellis L. Yochelson, *Charles Doolittle Walcott, Paleontologist* (Kent, Ohio & London: Kent State University Press, 1998), and *Walcott: Smithsonian Institution Secretary, Charles Doolittle Walcott* (Kent, Ohio & London: Kent State University Press, 2001). Yochelson describes in more detail Walcott's contribution to the geochronological debate in two papers: ' "Geologic time" as calculated by C.D. Walcott', *Earth Sciences History* **8**, part 2 (1989), 150–158, and that co-authored with Cherry Lewis detailed above.

William Johnston Sollas' major contributions were 'The age of the Earth', *Nature* **51** (1895), 533, and 'Anniversary address of the President: position of geology among the sciences; on time considered in relation to geological events and to the development of the organic world; the rigidity of the Earth and the age of the oceans', *Proceedings of the Geological Society* **65** (1909), lxxxi–cxxiv. The latter was used as Chapter 1 in his book *The Age of the Earth* (London: T. Fisher Unwin, 1905). For more on John Joly see P. N. Wyse Jackson, 'John Joly (1857–1933) and his determinations of the age of the Earth', in Lewis and Knell, *The Age of the Earth* (2001), pp. 107–119. His relevant publications were his paper 'The age of the Earth', *Philosophical Magazine*, Series 6, **22** (1911), 357–380 [Reprinted in the *Annual Report of the Smithsonian Institution* (for 1911), 271–293 (1912)], and Chapter 1 in his book *Birth-time of the World and Other Scientific Essays* (London: T. Fisher Unwin, 1915) which contained the transcript of his lecture delivered on 6 February 1914 to the Royal Dublin Society.

Table 10.1 is largely derived from Charles Schuchert, 'Geochronology, or the age of the Earth on the basis of sediments and life', in Adolph Knopf, *Physics of the Earth IV. The Age of the Earth* (Washington: Bulletin of the National Research Council, Number 80, 1931), pp. 10–64.

Chapter 11. Thermodynamics and the cooling Earth revisited

An almost contemporary biography of William Thomson is that by Andrew Grey, *Lord Kelvin: An Account of His Life and Work* (1908). A more recent account is given by Crosbie Smith and M. Norton Wise, *Energy and Empire: A Biographical Study of Lord Kelvin* (Cambridge University Press, 1989). A more popular treatment is given by Mark McCartney, 'William Thomson: king of Victorian physics', *Physics World* **15**, part 12 (2002), 25–29, which is an abridged version of his

'William Thomson, Lord Kelvin 1824–1907', in M. McCartney and A. Whitaker (eds.), *Physicists of Ireland: Passion and Precision* (Bristol and Philadelphia: Institute of Physics Publishers, 2002), pp. 116–125. Thomson's yacht the *Lalla Rookh* is described and illustrated in A. L. Rice, *British Oceanographic Vessels 1800–1950* (London: The Ray Society, 1986), a fascinating compendium of information on those ships that were the workhorses of oceanographic research.

The most comprehensive account of William Thomson's geochronological research is that by Joe D. Burchfield, *Lord Kelvin and the Age of the Earth*, 2nd edn (University of Chicago Press, 1990), which contains an exhaustive bibliography. Burchfield discusses in detail the methods used by Thomson and the counter arguments to his work, which are also found in Stephen G. Brush, 'Finding the age of the Earth by physics or by faith', *Journal of Geological Education* **30** (1982), 34–58. Arthur Stinner discusses the mathematical calculations used by Thomson in his geochronological papers in 'Calculating the age of the Earth and the Sun', *Physics Education* **37**, part 4 (2002), 296–305 from where Figure 11.2 is taken. In *Great Feuds in Science: Ten of the Liveliest Disputes Ever* (New York & Chichester: John Wiley & Sons, 1998) Hal Hellman recalls the debates between Thomson and the geologists; this essay contains the quotation of Mark Twain. A longer quotation from Twain appears in Burchfield, *Lord Kelvin* (1990), p. ix.

Thomson's most important papers on the age of the Sun and Earth were, firstly, those relating to the age of the Sun: 'On the mechanical energies of the Solar System', *Philosophical Magazine* **8** (1854), 409–430; 'On the age of the Sun's heat', *Macmillan's Magazine* **5** (1862), 288–393; 'On the Sun's heat', *Proceedings of the Royal Institution* **12** (1889), 1–12; secondly, on the cooling rate of the Earth: 'On the secular cooling of the Earth', *Transactions of the Royal Society of Edinburgh* **23** (1862), 157–170; and was followed by 'On Geological Time', *Transactions of the Geological Society of Glasgow* **3**, part 1 (1868), 1–28, and 'The age of the Earth', *Nature* **51** (1895), 438–440; and thirdly, on tidal friction and the Earth's rotation and shape: 'On Geological Time', *Transactions of the Geological Society of Glasgow* **3**, part 1 (1868), 1–28. His final paper contribution to the geochronological debate was 'The age of the earth as an abode fitted for life', *Annual Report of the Smithsonian Institution* for 1897 (1898), 337–357. [Reprinted in *Journal of the Transactions of the Victoria Institution* **31** (1899), 11–34, and in the *London, Edinburgh, and Dublin Philosophical Magazine and Journal of Science*, Series 5, **57** (1899), 66–90].

A critical examination of Clarence King's work is given in K. R. Aalto, 'Clarence King's geology', *Earth Sciences History* **23**, part 1 (2004), 9–31. Thomson's reaction to King's 1893 paper is summarised in Ellis L. Yochelson and Cherry Lewis, 'The age of the Earth in the United States (1891–1931): from the geological viewpoint', in Lewis and Knell *The Age of the Earth* (2001) pp. 139–155, while John Perry's objections are cogently analysed in a paper in the same volume by Brian Shipley 'Had Lord Kelvin a right?': John Perry, natural selection and the age of the Earth, 1894–1895', pp. 91–105.

Rutherford's recollection of his encounter with Kelvin at the Royal Institution in 1904 is documented in the biography by Arthur S. Eve, *Rutherford* (New York: Macmillan, 1939), p. 107, and also in Burchfield, *Lord Kelvin* (2001), p. 164 where it is discussed further.

James Blaylock's novel whose title refers to Kelvin is *Lord Kelvin's Machine* (Wisconsin: Arkham House, 1992).

Chapter 12. Oceanic salination reconsidered

Further information on Joly can be found in John R. Nudds, 'The life and work of John Joly (1857–1933)', *Irish Journal of Earth Sciences* 8 (1986), 81–94, and in P. N. Wyse Jackson, 'A man of invention: John Joly (1857–1933), engineer, physicist and geologist', in David S. Scott (ed.) *Treasures of the Mind: A Trinity College Dublin Quatercentenary Exhibition* (London: Sothebys, 1992), pp. 86–96; 158–160.

Some of Joly's poetry including that on *Oldhamia* was published long after his death in J. R. Nudds (ed.) *Upon Sweet Mountains: A Selection of Poetry by John Joly F.R.S.* (Dublin: Trinity Closet Press, 1983).

Joly's sodium method is dealt with in more detail in P. N. Wyse Jackson, 'John Joly (1857–1933) and his determinations of the age of the Earth', in Lewis and Knell, *The Age of the Earth* (2001), pp. 107–119, in which a comprehensive bibliography may be consulted. Joly's major paper on the sodium method of dating the Earth was 'An estimate of the geological age of the Earth', *Scientific Transactions of the Royal Dublin Society* 7, (1899), 23–66 [Reprinted in *Annual Report of the Smithsonian Institution* for 1899, (1901), 247–288]. Further publications by Joly that largely explained and defended his 1899 work were 'A fractionating rain-gauge', *Scientific Proceedings of the Royal Dublin Society* 9 (1900), 283–288; 'Geological Age of the Earth', *Geological Magazine*, New Series, 7 (1900), 220–225 [Reprinted in *Nature* 62 (1900), 235–237]; 'On Geological Age of the Earth', *Report of the British Association for the Advancement of Science, Bradford 1899*, Section C3 (1900), 369–379; 'Some experiments on denudation in fresh and salt water', *ibid*, 731–732 [Also published in the proceedings of the 8th International Geological Congress, Paris 1901, and enlarged in the *Proceedings of the Royal Irish Academy* 24 A (1902), 21–32.]; 'The circulation of salt and geological time', *Chemical News* 83 (1901), 301–303; 'The circulation of salt and geological time', *Geological Magazine*, New Series, 8 (1901), 354–350; and *Birth-time of the World and Other Scientific Essays* (London: T. Fisher Unwin, 1915). In 1930 in *Surface History of the Earth*,

2nd edn (Oxford University Press, 1930) Joly accepted some of A. C. Lane's arguments that suggested 300 million years was a better estimate for his sodium method.

Various publications by a number of English commentators on the sodium method and chemical denudation are worthwhile consulting: T. M. Reade, 'President's Address', *Proceedings of the Liverpool Geological Society* **3**, part 3, (1876), 211–235; T. M. Reade, *Chemical Denudation in Relation to Geological Time* (London: Daniel Dogue, 1879); and T. M. Reade, 'Measurement of geological time', *Geological Magazine*, New Series, **10** (1893), 97–100; W. J. Sollas, 'Anniversary address of the President: position of geology among the sciences; on time considered in relation to geological events and to the development of the organic world; the rigidity of the Earth and the age of the oceans', *Proceedings of the Geological Society* **65** (1909), lxxxi–cxxiv. Osmund Fisher's review of Joly's 1899 paper makes interesting reading: 'Review of "An estimate of the Geological Age of the Earth", by John Joly', *Geological Magazine*, New Series, **7** (1900), 124–132.

Reaction from the United States came in G. F. Becker, 'Reflections on J. Joly's method of determining the ocean's age', *Science* **31** (1910), 509–512; F. W. Clarke, 'A preliminary study of chemical denudation', *Smithsonian Miscellaneous Collections* **56**, part 5 (1910), 1–19; G. F. Becker, 'The age of the Earth. *Smithsonian Miscellaneous Collections* **56**, part 6 (1910), 1–28; H. S. Shelton, 'The age of the earth and the saltness of the sea', *Journal of Geology* **18** (1910), 190–193; and A. C. Lane, 'The Earth's age by sodium accumulation', *American Journal of Science*, Series 5, **17** (1929), 342–346.

Joly's work was finally discredited by the third decade of the twentieth century. See A. Harker, 'Some remarks on geology in relation to the exact sciences, with an excursus on geological time', *Proceedings of the Yorkshire Geological Society* **19** (1914), 1–13; J. Barrell, 'Rhythms and the measurements of geologic time', *Geological Society of America Bulletin* **28** (1917), 745–904; J. W. Gregory, 'The age of the Earth', *Nature* **108** (1921), 283–284; T. C. Chamberlin, 'The age of the earth from the geological viewpoint', *Proceedings of the American Philosophical Society* **61**, part 4 (1922), 247–271; and Arthur Holmes, 'Estimates of geological time, with special reference to thorium minerals and uranium haloes', *Philosophical Magazine*, Series 7, **1** (1926), 1055–1074.

For a modern assessment of the composition of early oceans see L. P. Knauth, 'Salinity history of the Earth's early ocean', *Nature* **395** (1998), 554–555.

Chapter 13. Radioactivity: invisible geochronometers

The most comprehensive account of the history and modern use of radioactivity in dating rocks is that by G. Brent Dalrymple, *The Age of the Earth* (Stanford

University Press, 1991) from where (p. 80) I have taken the half-lives of the decay series listed. Stephen G. Brush's 'Finding the age of the Earth by physics or by faith?', *Journal of Geological Education* **30** (1982), 34–58, contains an in-depth account of the radioactive methodologies used to derive dates from rocks, minerals and other geological matter.

The contribution of Bertram Boltwood is discussed in detail by Lawrence Badash, 'Rutherford, Boltwood, and the age of the Earth; the origin of radioactive dating techniques', *Proceedings of the American Philosophical Society* **112**, part 3 (1968), 157–169; their letters are reproduced in L. Badash (ed.), *Rutherford and Boltwood: Letters on Radioactivity* (New Haven: Yale University Press, 1969). A useful and brief account of the development of the uranium decay series is found in G. M. Henderson, 'One hundred years ago: the birth of uranium-series science', in B. Bourdon, G. M. Henderson, C. C. Lundstrom and S. O. P. Turner (eds.) *Uranium-Series Geochemistry*. Reviews in Mineralogy & Geochemistry **52** (Geochemical Society and Mineralogical Society of America, 2003), pp. v–x.

John Joly suggested that heat invalidated Thomson's geochronology in the note 'Radium and the geological age of the Earth', *Nature* **68**, (1903), 526. Joly's most important works on radioactivity were summarised in his *Radioactivity and Geology* (London: Archibald Constable & Co. Ltd, 1909). The history of the Irish Radium Institute is given in J. Joly, 'History of the Irish Radium Institute (1914–1930)', *Royal Dublin Society Bicentenary Souvenir 1731–1931* (Royal Dublin Society, Dublin, 1931) pp. 23–32; and in David J. Murnaghan, 'History of radium therapy in Ireland: the 'Dublin Method' and the Irish Radium Institute', *Journal of the Irish Colleges of Physicians and Surgeons* **17**, part 4 (1988), 174–176.

Over ten or so years Joly produced a large volume of work on pleochroic halos which is best read in J. Joly, 'Pleochroic halos', *Philosophical Magazine*, Series 6, **13** (1907), 381–383; J. Joly and E. Rutherford, 'The age of pleochroic haloes', *Philosophical Magazine*, Series 6, **25** (1913), 644–657; and J. Joly, 'The genesis of pleochroic haloes', *Philosophical Transactions of the Royal Society* **217** (1917), 51–79. The validity of Joly's findings was disputed by D. E. Kerr-Lawson, Pleochroic haloes in biotite from near Murray Bay, Province of Quebec', *Toronto University Studies*, Geology Series, **24** (1927), 54–70, and by Franz Lotze, 'Pleochroic haloes and the age of the Earth', *Nature* **121** (1938), 90.

I owe a great debt of gratitude to Cherry Lewis whose wonderful research and writings on Arthur Holmes have provided me with much of the information on the man and his work given here. In 2000 she published the biographical study *The Dating Game: One Man's Search for the Age of the Earth* (Cambridge University Press, 2000), and followed this with 'Arthur Holmes' vision of a geological timescale', pp. 121–138, in Lewis and Knell, *The Age of the Earth* (2001), pp. 121–138. She discusses the development of his ideas on mantle convection currents in 'Arthur Holmes' unifying theory: from radioactivity to continental drift', in D. R. Oldroyd (ed.), *The Earth Inside and Out: Some Major Contributions to Geology in the Twentieth Century* (London: Geological Society of London Special Publication 192, 2002), pp. 167–183.

A bibliography of Arthur Holmes' work is given in F. H. Stewart, 'Arthur Holmes', *Quarterly Journal of the Geological Society of London* **120** (1964), 3–11.

Holmes' first paper on radioactivity was 'The association of lead with uranium in rock-minerals and its application to the measurement of geological time', *Proceedings of the Royal Society* **A85** (1911), 248–256; and he followed this with others including these in which he constructed a geological timescale: 'Radioactivity and the measurement of geological time', *Proceedings of the Geologists' Association* **26**, part 5 (1915), 289–309; 'The measurement of geological time', *Nature* **135** (1935), 680–685; 'The construction of a geological time-scale', *Transactions of the Geological Society of Glasgow* **21**, part 1 (1947) 117–152; 'A revised geological time-scale', *Transactions of the Edinburgh Geological Society* **17**, part 3 (1960), 183–216. Holmes also wrote a small popular book that is well worth reading: *The Age of the Earth* (London & New York: Harper & Brothers, 1913). He later rewrote it (London: Ernest Benn, 1927) and it went through several editions right up to 1937 when it was published in London by Thomas Nelson & Sons.

The 2004 geological timescale is modified from F. M. Gradstein, J. G. Ogg, A. G. Smith, W. Bleeker and L. J. Lourens, 'A new geologic time scale, with special reference to Precambrian and Neogene', *Episodes* **27**, part 2 (2004), 83–100.

Holmes' model for dating using the isotopic abundance of lead appeared in 'An estimate of the age of the Earth', *Nature* **157** (1946), 680–684. This together with the papers by Gerling (1942) and Houtermans (1946) is reproduced in C. T. Harper (ed.) *Geochronology: Radiometric Dating of Rocks and Minerals* (Stroudsberg PA: Dowden, Hutchinson & Ross, Inc., 1973). Figure 13.6 is taken from A. Holmes, 'An estimate of the age of the Earth', *Geological Magazine* **84** (1947), 123–126.

Primeval lead was introduced in 1939 by Alfred Nier, 'The isotopic composition of radiogenic leads and the measurement of geological time. II', *Physical Review* **55** (1939), 153–163. Nier recalled his scientific work in 'Some reminiscences of isotopes, geochronology, and mass spectrometry', *Annual Review of Earth and Planetary Sciences* **9** (1981), 1–18. A biographical memoir by John H. Reynolds appears in *Biographical Memoirs, National Academy of Sciences* **74** (1998), 244–265.

Chapter 14. The Universal problem and Duck Soup

A fascinating account of Clair Patterson's work on the question of the age of the Earth, and his later environmental research, was recalled by him in interviews conducted on 5, 6 and 9 March 1995 by Shirley K. Cohen for the Caltech Archives Oral History Project. Parts of these were published in 1997 under the title 'Duck soup and lead' in the Caltech journal *Engineering & Science* **60**, part 1 (1997), 21–31. It is a great shame that few similar oral histories exist for other scientists elsewhere – they often provide insights infrequently touched on in obituaries. A comprehensive account of Patterson's work is given by his colleague and contemporary George Tilton in a memoir prepared for the National Academy of Sciences, Washington. This includes a selected bibliography: *Biographical Memoirs, National Academy of Sciences*, **74** (1998), 266–287. Appropriately, this follows the memoir for Al Nier. The quotation of Patterson's thoughts on his predecessors was published in the Clair C. Patterson Special Issue of *Geochimica et Cosmochimica Acta* **58**, part 15 (1994), 3141–3143.

The publications by Patterson and his colleagues on research that led to their determination of the age of the Earth were: C. C. Patterson, 'The isotopic composition of meteoritic, basaltic and oceanic leads, and the age of the Earth', *Proceedings of the Conference on Nuclear Processes in Geologic Settings*, Williams Bay, Wisconsin, September 21–23 (1953), 36–40 (this was the first paper to suggest a date of 4,500 million years); C. Patterson, G. Tilton and M. Inghram, 'Abundances of uranium and the isotopes of lead in the Earth's crust and meteorites', *Bulletin of the Geological Society of America* **64**, number 12, part 2 (1953), 1461; C. Patterson, H. Brown, G. Tilton and M. Inghram, 'Concentration of uranium and lead in the isotopic composition of lead in meteoritic material', *Physical Review* **92** (1953), 1234–1235; C. Patterson, 'The Pb^{207}/Pb^{206} ages of some stone meteorites', *Geochima et Cosmochimica Acta* **7** (1955), 151–153; C. Patterson, 'Age of meteorites and the earth', *Geochima et Cosmochimica Acta* **10** (1956), 230–237 (which many regard as the classic paper in this series).

Stephen G. Brush has written extensively about the dating of the Earth and Universe, and his treatment of the last century is best found in three papers 'Finding the age of the Earth by physics or by faith?', *Journal of Geological Education* **30** (1982), 34–58; 'The age of the Earth in the twentieth century', *Earth Sciences History* **8**, number 2 (1989), 170–182; and 'Is the Earth too old? The impact of geochronology on cosmology: 1929–1952', in Lewis and Knell, *The Age of the Earth* (2001), pp. 157–175 London, 2001). Norriss S. Hetherington's paper 'Geological time versus astronomical time: are scientific theories falsifiable', *Earth Sciences History* **8**, number 2 (1989), 167–169, deals with the same problem but in less detail.

I cannot stress enough the value of G. Brent Dalrymple's book *The Age of the Earth* (1991). It remains without question the best treatment of the history and methodology of geochronological dating methods particularly those used in the last one hundred years. He gives a synopsis of the book in G. B. Dalrymple, 'The age of the Earth in the twentieth century: a problem (mostly) solved', in Lewis and Knell, *The Age of the Earth* (2001) pp. 205–221.

Index

Page number in italic denotes a figure.

Académie Royale des Sciences, Paris 108
Agassiz, Louis 55, 147, 148
age of the Earth 258
    based on cooling rates 111–117, 201–203
    based on meteorites 255–256, 257, 258
    based on salinity of the oceans 219–221
    based on sedimentation rates 177–179, 188,
        189–196
    based on the age of the Sun 200–201
    based on tidal friction measurements 203
    biblical estimates 14–27
    compared to that of the Universe
        250–252
Airy, George Biddell 174
Alberti, Friedrich August von 143
Allen, Thomas 14, 18, 43
American Association for the Advancement
    of Science 190
Ammonites, snakestones 158, *159*
Anaximander 8
Anaximenes 9
Anning, Joseph 139
Anning, Mary 139
Arduino, Giovanni 78, 79–81, *80*, 135, 137
Aristotle 32
Arkell, William Joscelyn 164
Ashmole, Elias 50
Ashmolean Museum, Oxford 49, 57, 179
Assur-bani-pal, King of Assyria 4, *5*

Bakewell, Robert 129
Baird, John Logie 227
Balfour, John Hutton 99
Barnes, Howard T. 230
Barrande, Joachim 164
Barrell, Joseph 195, 224
Barus, Carl 202
Bayeux Tapestry 58
Becker, Andrew 61

Becker, George Ferdinand 62, 206, 219,
    223, 237
Becquerel, Antoine Henri 233
    discovery of radioactivity 228
Berger, Jean-François 96, 99
Bergman, Torbern Olof 78, 81–82
Bergomensis, Philip 17
Bible
    Bishop's 29
    Geneva 29
    King James 13, 14, 30, 31
    Massoretic text 14
    Septuagint Greek text 14
    Vulgate, Latin 14
Black, Joseph 88
Bonaparte, Napoleon 171
Bone, Charles Richard
    illustrator of fossils 162
Boltwood, Bertram Borden 236, 238
    radiometric dating using radioactive
        element isotope ratios 237
Boyle, Robert 72
Braun, Karl Ferdinand 227
Brice, William 23
British Association for the Advancement of
    Science 128, 178, 179, 205, 208, 216,
    222, 251
Brongniart, Alexandre 160
Brookes, Richard 51
Broughton, Hugh 17
Brown, Harrison Scott 252, 253
Brush, Stephen 250
Bryson, Bill 253
Buch, Christian Leopold von 84, 138
Buckland, William 51, 119, 129, 146, 147
Buckman, Sydney Savory 164
Buffon, Georges-Louis Leclerc, Comte de
    106–118, *107*, 200, 201, 233
    'Buffon's needle problem' 108

Buffon, Georges-Louis Leclerc,
     Comte de (cont.)
  Buffonet, Georges-Louis-Marie
       Leclerc, son 117
  builds forge for heating experiments 112, *113*
  death 117
  Estate at Montbard near Dijon 108,
       110–111, 112, *112*
  Histoire naturelle 109–111
  on the age of the Earth 111–117
  Théorie de la Terre 38
Bullinger, Heinrich 17
Burnet, Thomas 33, *33*, 38, 56, 79, 88
  his theory of the Earth 38–42, *40*
  Telluris Theoria Sacra 39–41, *40*
Butcher, Norman 104

California Institute of Technology (Caltech)
     253, 255, 259
Capellini, Giovanni 67
Carroll, Lewis 50
Challoner, Luke 21
Chamberlin, Thomas Chrowder 204, 224
Chambers, Robert 169
Charles I 15, 19
Charpentier, Jean de 147
Clarke, Frank Wigglesworth 223
Clausius, Rudolph Julius Emanuel
  Second Law of Thermodynamics 200
Clerk, George (Clerk Maxwell) 89
Clerk, John 99
Clerk, John (Lord Eldin) 100, 101
Clodd, Edward 162
Cloud, Preston Ercelle, junior 133
coelacanth, primitive fish 156
Colonna, Fabio 157
Conybeare, William 99, 138, 140
Courtenay-Latimer, Marjorie 156
Cooper's Chronology 16, *16*
Creation beliefs 1
  Chaldean and Babylonian 3–5
  Chinese and Japanese 7–8
  Egyptian 2–4
  Greek 8–9
  Indian (Vetic) 5–6
     Buddist 6
     Hindu 6
  Mayan 10
  Pacific 10–11
  Scandinavian 9–10

Croll, James 187
Cromwell, Oliver 16, 19, 20
Crookes, William 227
Curie, Marie 228, 233
  discovery of radium 228
Curie, Pierre 228, 229, 230
  discovery of radium 228
  on radium as a heat source 229, 230
Cuvier, Georges 106, 160, 168, 169
cyclical nature of Earth history 74, 92

Dana, James Dwight 134
Darwin, Charles 8, 73, 162, 164, 181, *182*,
     189, 200
  as a geologist 185–187
  on the erosion of the Weald 180–183
  opinion of William Thomson 207
  theory of evolution 169–170
Darwin, George Howard 204
Davidson, Thomas 164
da Vinci, Leonardo 64, 69, 157
Davy, Humphry 96
Debierne, André Louis 228, 233
de Koninck, Guillaume 164
De la Beche, Henry Thomas 121, 140
  on colouring geological maps 151
Descartes, Réne 32, 34–36, 39, 64
  theory of the Earth 35–36
  Principia philosophiae 35
  skull 35
de Sitter, Willem 251
Desmarest, Nicholas 93, 96
  on volcanic origin of basalt 93
Desnoyers, Jules Pierre François
     Stanislas 146
d'Halloy, Jean Baptiste Julien, d'Omalius 138
dinosaur
  Megalosaurus 51, 52
Dixon, Henry Horatio 211
d'Orbigny, Alicide Dessalines 165
Dove, Jonathan 18
Drake, Roger 17
Drury, Susanna 96
Dublin Society, later Royal
     Dublin Society 96

Edmunds, John 54
Einstein, Albert 50
Emmons, Samuel Franklin 134
Empedocles 9

Farey, John 125, 181
Fell, John 30
Fisher, Osmund 204, 221
Fitton, William Henry
    Geological map of Dublin 151
Fitzgerald, George Francis 214
Fletcher, Arnold Lockhart 239
Forbes, Edward 58, 146
fossils 154–170
    and evolutionary studies 168–170
    as zone fossils 165–168
    folklore 157–158, *159*
    'living fossil' 156
    nature of 69
    oldest on Earth 154
    Oldhamia radiata, Cambrian trace
        fossil 217
    palaeontology: study of fossils 154
    textbooks on 165
    use in correlating rocks 165–168
    use in dating rocks 73, 159
Franklin, Benjamin 109
Fuller, John 30
Fullerton, James 21

Geiger, Hans 228
Geikie, Archibald 75, 103, 208
    on denudation rates 189
    on the age of rocks in Wales 149–150
Geikie, James 204
geological column *ii*
    absolute dates for 241
    naming its sub-divisions 133–135, 136
    naming the geological periods 135–150
geological maps
    colouring of 150–152
Geological Society of America 195, 249, 257
Geological Society of France 146
Geological Society of Glasgow 202, 203
Geological Society of London 103,
        119, 127, 133, 140, 145, 146,
        147, 148, 149, 167, 178, 179,
        186, 202, 216, 249
    Wollaston Medal 128, 147, 179, 186, 249
Geological Survey of Canada 121, 167
Geological Survey of England and Wales 121,
        146, 149
Geological Survey of India 121
Geological Survey of Ireland 119
geological time 131–133

geological timescale 241
Gerling, Erik Karlovich 248
Gervais, Paul 146
Giant's Causeway 97
    origin of basalt of 96–99
Gibson, Edmund 54
Glen Tilt, Scotland 99, *100*, 104
Grace, W. G. 219
Greenough, George Bellas 128
    on colouring geological maps 151
Goodchild, John George 194
Gordon, George 149
Gould, Stephen Jay 23, 42
Grabau, William 166
Grand Canyon, Colorado 172
Gregory, John Walter 224
Griffith, Richard 164
Gualterus, Rodolphus 17
Guettard, Jean-Étienne
    geological map of France 151

Hales, William 28
Hall, James 67
Hall, Sir James 88, 96, 98
Halley, Edmond 58–62, *59*, 65, 172, 218
    comet 58, 59
    diving bell 60
    on salinity method to age the Earth *61*,
        61–62, 218
    on the hydrological cycle 61
    visits St Helena 61
Halstead, Beverly 51
Hamilton, Revd. William
    on volcanic origin of Giant's
        Causeway 97
Harker, Alfred 224
Haughton, Samuel 189, 193, 203, 216
Hawkins, Benjamin Waterhouse 139
Henslow, John Stephens 183, 186
    mentor of Charles Darwin 183
Heraclitus 9
Herodotus 9
    on sedimentation rates as age indicator
        171–172
Hervey, Augustus, Earl of Bristol 96
Hesiodus 8
Hicks, Henry
    on the age of rocks in Wales 149–150
Hobson, Bernard 167, 194
Holland, Charles Hepworth 122

Holmes, Arthur 187, 224, 238, *238*, 241–244, 249
  geological timescale for the geological column 231, 244–249, *247*
  isochron plots showing age of Earth 233, 248, *248*
Holmes–Houtermans Model 229, 248, 249, 255
Hooke, Robert 71–74, 88
  Discourse of Earthquakes 73, 88
  Micrographia 72, 88
  on fossils 73
  on strata 74
Hooker, Joseph Dalton 183
Horner, Leonard 28, 103
Hornes, M. 135
Houtermans, Friedrich Georg 248
Hubble, Edwin Powell 251
Humason, Milton La Salle 251
Humboldt, Alexander von 84, 137, 180
Hunt, Thomas Streey 67
Hutton, James 72, 86–92, *87*, 131, 176, 189
  at Siccar Point 101
  his theory of the Earth 89–92, *90*, 99–103
  house in Edinburgh 103
  locates and describes unconformities 100, 101, *102*
  locates veins in granite at Glen Tilt 99, *100*
  on the Isle of Arran 100–101
  reaction to his theory of the Earth 93–99
  Theory of the Earth, with Proofs and Illustrations 103, 104
Huxley, Thomas Henry 202

Inghram, Mark 255
International Geological Congress 67
International Union of Geological Sciences 122
Isle of Arran, Scotland 100–101

James I 29
Jameson, Robert 84, 100, 185
Jardin des Plantes, Paris 105
Jardin du Roi, Paris 105, 108–109
Jeans, James Hopwood
  age of the stars 251
Jefferson, Thomas 109
John, Archduke of Austria 180
John Paul II, Pope 66
Johnson, Andrew, US President 197

Joly, John 211, *212*, 213–217, 229, 230, 237
  establishes Radium Institute, Dublin 229, 231
  on the age of the Earth 217–226, 239–241
  on sedimentation rates 194
  on pleochroic halos as geological clocks *239*, 239–241
  poetry, on *Oldhamia radiata*, Cambrian trace fossil 217
  radioactivity and geology 230, 242
  reaction to Joly's sodium method 221–223
  oceanic sodium method for estimating age of the Earth 213, 219–221
Jukes, Joseph Beete 119, 147

Kelvin, Lord (*see* Thomson, William)
Kerr-Lawson, D. E. 240
King, Clarence 202, 222
Kircher, Athanasius 32, 36–38
  Mundus subterraneus 37
  theory of the Earth 37–38
Kirwan, Richard 93–95, 98, 103, 140
Klaproth, Martin 228

Laborde, Albert 229, 231
Lacépède, Bernard Germain Étienne de la Ville, Comte de 109
Lamarck, Jean Baptiste Antoine de Monet de
  ideas on evolution 168–169
Lane, Alfred Church 225
Lapworth, Charles 144
Lawrence, T. E. (Lawrence of Arabia) 49
lead isotopes 254–255
  primeval lead 229, 248, 249, 254
  ratios in meteorites 255
Lehmann, Johann Gottlob 78–79
Leicester, Robert 29
Lewis, Cherry 249
Lhwyd, Edward 47–58, *48*, 65, 158
  geological collections 48, 52–54
  on fossils 52, *53*, 158
  on the age of the Earth 56–57
  term 'Celtic' 54, 55
Libby, Willard Frank 252
  develops carbon-14 dating 252
Lightfoot, John 27–28, 31
Linnaeus, Carl 81, 109, 159
Linnean Society of London 170
Lister, Martin 158
Liverpool Geological Society 193

Llanberis, north Wales 52, 55, *56*, 58
Lloyd, William 30, 31
Locke, John 46
Lotze, Franz 240
Louis XIII 105
Louis XVI 105, 117
Lowry, James Wilson
    illustrator of fossils 162
Luc, Jean-André de 95
Lucretius, Carus Titus 1
Lyell, Charles 119, *121*, 126, 128–131, 133,
        142, 146, 147, 176, 183, 187, 201
    Elements of Geology 130
    Principles of Geology 129, 131, *132*, 186
    Uniformitarianism 130–131

M'Coy, Frederick 164
McGee, William John 192, 194
Maillet, Benoît de 63–65, 78
    theory of the Earth 64
Mallet, Robert 75
Manhattan Project 252, 253
Marshall, Benjamin 30
Martin, William
    on fossils 159
meteorites 255, *256*
    age of Canyon Diablo meteorite
        255–256, 257
    age of Henbury meteorite 258
    analysed by Clair Patterson 257
Michell, John 75, 124
Mills, Abraham 96
Milner, John 23
Mohs, Frederick
    on hardness of minerals 176
Montano, Benito Arias 17
Moore, Raymond C. 167
Morell, Jack 185
Mount Etna, Sicily 38
Murchison, Roderick Impey 119, *120*, 143,
        144, 145–146, 149
Murray, John, 4th Duke of Atholl 99
Murray, Sir John 220
Murray, George, 6th Duke of Atholl 99

Napoleon Bonaparte 59
Naumann, C. F. 135
Nelson, Horatio 171
Newton, Isaac 46, 52, 59, 60, 71, 111, 259
Nicholson, Henry Alleyne 165

Nier, Alfred Otto Carl 246, 249
    work with mass spectrometers 246
    primeval lead 254
Nisbit, William 15, 18
oceans *218*
Oppel, Albert 166
Owen, Richard 164

Parkinson, John
    on fossils 159
Parry, David 54
Pasteur, Louis 36
Patterson, Clair Cameron *253*,
        253–260
    age of the Earth based on meteorite
        studies 255–256, 257, 258, *258*
    meteorites analysed *256*, 257
Perry, John 205, 206
Peterborough, Dowager Countess
        19, 21
Phillips, John 126, 134, *177*, 177–179,
        183–185, 187, 189, 200
    on sedimentation rates and time
        179–180, 195
    on the erosion of the Weald
        180–183
    problem with sedimentation rates as a
        geochronometer 187–189
Phillips, William 140
Picasso, Pablo 110
Pictet, François Jules 165
Playfair, John 88, 98, 103
pleochroic halos *239*
    as geological clocks 239–241
    problems with their use as geological
        clocks 240
Pliny (the Elder) 157
Plot, Robert 49–52
    on fossils 51, 158
Pont, Mr 17
Porter, Roy 38
Portlock, Joseph Ellison 119
    on geological mapping 151
Poulton, Edward Bagnall 205
Pratt, John Henry 174
Price, Hugh 49
Ptah, Egyptian creator god 4, *4*
Pythagoras 9, 157

Quenstedt, Friedrich 166

radioactivity 228
  decay sequences of radioactive elements
      230, 232–233, *234*, 242
  heat source of the Earth 229–230
  isotopes 233
radiometric dating using radioactive element
      isotope ratios 237, 251
  Holmes–Houtermans Model 248–249
radium 233, *235*, 248
  discovery 228
  as a heat source 229, 230
  in cancer treatments 229, 231
Ramsay, Andrew Crombie 182, 184
Ravis, Christian 19
Ray, John 55, 56, 76
Reade, Thomas Mellard 193, 194, 204, 219
  on sedimentation and solution rates 193
Reynolds, Doris Livesey 244
Reynolds, John 257, 258
Richardson, William 97, 130
Röntgen, Wilhelm 227
  discovery of X-rays 227
Royal Dublin Society 195, 219
  Radium Institute 229, 231
Royal Geographical Society 146
Royal Institution, London 209
Royal Irish Academy 97
Royal Society of Edinburgh 86
Royal Society of London 18, 43, 57, 59, 72,
      94, 108, 140, 214, 216
Royal Zoological Society, London 110
Rudwick, Martin 45, 160, 169
Rutherford, Ernest 209, 229, *230*, 231–232,
      233, 239, 248, 249

sacred theories of the Earth 38–46
Saint-Fond, Barthélemi de 117
Saint-Hilaire, Geoffroy 106
Salisbury, Marquess of 205
Sandage, Allan Rex 252
Sarjeant, Bill 51
Scaliger, Joseph Justus 17, 24
Sedgwick, Adam 119, *120*, 131, 134, 143,
      144, 186
sedimentation rates
  as an indicator of Earth's age 172–173,
      176–196
  problem with sedimentation rates as a
      geochronometer 187–189, 195–196
Seymour, Webb 98

Shimer, Hervey Woodburn 166
Siccar Point, Scotland 86, *102*
Siddhartha Gautama (Buddha) 6
Shipley, Brian 206
Sloane, Hans 52
Smith, Adam 88
Smith, James Leonard Brierley 156
Smith, William 70, 126–128, *127*, 156,
      178, 179
  *A Delineation of the Strata of England
      and Wales* 127
  geological laws 128
  on colouring geological maps 151
  on fossils as stratigraphical indicators
      160–162, *161*, 165
Smith, William Campbell 242
Smyth, Louis Bouvier 216
Snowdon Lily 54, 55, 58
Soddy, Frederick 229, 233
sodium method for estimating age of the
      Earth 213
Sollas, William Johnson 187, 194, 215, 222
Steno, Nicolaus 66–71, *67*, 75, 88
  on fossils 69, 157
  on minerals 69, 73
  on the Tuscan landscape 67, *68*, 69
  Prodromus 67, 75
Stephens, Walter
  geological map of Dublin 151
Stevenson, Walter Clegg 231, 245
Strachey, John 76, *76*, 124
Strange, John 130
stratification 66
  Hooke on 74
  Michell on 75, 124
  Strachey on 76, *76*
stratigraphical charts *163*
stratigraphical laws 126, 128
stratigraphy 122–123
  development of a global scheme 77
  early stratigraphical schemes 123–125
Strutt, Robert John (Baron Rayleigh) 236, 243
Stukeley, William 77
Suess, Eduard
  The Face of the Earth 242
Swan, John 15

Tait, Peter Guthrie 201
Tennant, Smithson 228
Theophilus of Antioch 13

Thomson, Joseph John 229, 231
Thomson, William (Lord Kelvin) 184, 189,
    190, 197–207, *198*, 220, 242, 246
  coat of arms 206, *207*
  on cooling rates of the Earth 201–203,
    229, 230
  on the age of the Earth 200–203
  on the age of the Sun 200–201
  on tidal friction 203
  honours 206
  reactions to his age estimates 203–206,
    207–209, 229, 230
Tilton, George 254
Toulmin, George Hoggart 88
Trinity College Dublin 20, 52, 54, 97, 98, 128,
    177, 189, 194, 211, 214
Twain, Mark 206

unconformities 100
United States Geological Survey 167, 192
Universe
  age 250–252
University of Chicago 252, 255
  Argonne National Laboratory 255
Upham, Warren 192
Urey, Harold Clayton 252
Ussher, James 13, 16, *17*, 19–27, 30, 31,
    54, 56
  Annals of the World 22
  Annales veteris testamenti 22, *25*

Venetz-Sitten, Ignace 147
Verneuil, Philippe Édouard Poulletier de 145

Victoria, Queen 197
Vivarès, François 96

Walcott, Charles Doolittle
  on sedimentation rates in USA
    190–192, *191*
Wallace, Alfred Russel 170, 193
Waller, Richard 73
Wasserburg, Gerard 253, 255, 258
Watt, James 89
Weald, Sussex *181*
  Charles Darwin and John Phillips on
    erosion of 180–183
Werner, Abraham Gottlob 77, 78, 82–84, 93
Wesley, John 96
Westminster Abbey 60, 131, 207
Whiston, William 33, 45–46, 56
Whitehurst, John 96
Willet, Andrew 17
Williams, Henry Shaler 141
Winchell, Alexander 141
Wood, Searles Valentine, Snr 178, 179
Woodward, John 33, 42, 56, 57, 79, 119
  theory of the Earth 42–45
Wren, Christopher 49, 71

Xenophanes 9

Yochelson, Ellis 190
Young, George 87

Zittel, Karl Alfred von 67, 165
Zoroaster 29